I0092077

The Archaeology of Montebello Islands, North-West Australia

Late Quaternary foragers on an arid coastline

Peter Veth
Ken Aplin
Lynley Wallis
Tiina Manne
Tim Pulsford
Elizabeth White
Alan Chappell

BAR International Series 1668
2007

Published in 2016 by
BAR Publishing, Oxford

BAR International Series 1668

The Archaeology of Montebello Islands, North-West Australia

ISBN 978 1 4073 0103 7

© The authors individually and the Publisher 2007

The authors' moral rights under the 1988 UK Copyright,
Designs and Patents Act are hereby expressly asserted.

All rights reserved. No part of this work may be copied, reproduced, stored,
sold, distributed, scanned, saved in any form of digital format or transmitted
in any form digitally, without the written permission of the Publisher.

BAR Publishing is the trading name of British Archaeological Reports (Oxford) Ltd.
British Archaeological Reports was first incorporated in 1974 to publish the BAR
Series, International and British. In 1992 Hadrian Books Ltd became part of the BAR
group. This volume was originally published by Archaeopress in conjunction with
British Archaeological Reports (Oxford) Ltd / Hadrian Books Ltd, the Series principal
publisher, in 2007. This present volume is published by BAR Publishing, 2016.

Printed in England

BAR
PUBLISHING

BAR titles are available from:

        BAR Publishing
        122 Banbury Rd, Oxford, OX2 7BP, UK
EMAIL   info@barpublishing.com
PHONE   +44 (0)1865 310431
FAX   +44 (0)1865 316916
        www.barpublishing.com

Montebello Islands Archaeology: Late Quaternary Foragers on an Arid Coastline

Peter Veth[1], Ken Aplin [2], Lynley A. Wallis[3], Tiina Manne[4],
Tim Pulsford[5], Elizabeth White[6] and Alan Chappell[7]

1. Australian Institute of Aboriginal and Torres Strait Islander Studies, Canberra, ACT
2. Australian National Wildlife Collection, C.S.I.R.O. Canberra, ACT
3. Department of Archaeology, Flinders University, Adelaide, SA
4. Department of Anthropology, University of Arizona, Tucson, Arizona
5. Environmental Protection Agency, Bundaberg, QLD
6. Jo McDonald Cultural Heritage Management Pty Ltd, Sydney, NSW
7. Advanced Analytical Centre, James Cook University, Townsville, QLD

Abstract

The Montebello Islands are a cluster of small, low relief land masses, comprised of ancient limestone, with skeletal soils, sparse vegetation and shifting sand bodies. They lie some 80 km from the coastline, representing far flung 'high points' on the once extensive arid coastal plains of north-west Australia. Barrow Island lies between the mainland and the islands. More famous as the first nuclear testing site used by the British in the 1950s and the location of the first known shipwreck off the Australian coast, (the *Tryal* in 1622), the Montebello Islands represent a unique configuration of terrestrial and marine ecosystems, with islands either connected or separated by mangrove flats, sand spits, shallow channels and limestone pavements and coral formations. The high biodiversity of the marine zone (as opposed to the now depauperate terrestrial fauna of the islands) has been recognised by its listing as a marine park.

This paper reports on archaeological analysis carried out on assemblages recovered from two stratified cave sites on Campbell Island in the Montebello group in northwest Australia. These sites provide unique insights into human responses to the drowning of the extensive arid plains of north-west Australia following the Last Glacial Maximum. Rich faunal assemblages have been recovered which date to the period 30 000 – 7 000 BP as the local environmental context changed in response to the post-glacial marine transgression.

Field surveys and excavations were carried out over two field seasons between 1992-1994 and involved a team of archaeologists, field assistants and support crew. Of particular note were field contributions made by Emeritus Professor Jim Allen, Ms Jill Allen and Drs Joe Dortch, Sue O'Connor and Bruce Veitch. Analyses of the surprisingly rich marine and terrestrial faunas (both anthropogenic and natural), sediments and artefacts recovered from the excavations have continued for approximately 10 years and included a range of postgraduate students and specialists (as cited below). Of special note with respect to advice on the faunas we wish to acknowledge Dr Alex Baynes (WA Museum), Dr Peter Kendrick (CALM), Professor Mary Steiner (University of Arizona), Dr Shirley Slack-Smith (WA Museum), Dr John Scanlon and Dr David Bellwood (James Cook University),

The archaeological record shows that occupation of the Montebello caves between approximately 30 000 and 10 000 BP was sparse and episodic in nature. This is similar to the pattern from similarly-aged stratified sites at North West Cape (e.g. Morse 1999). All of the hard rock stone artefacts which have been recovered from the Montebello Islands during this period must have been transported from supply zones on the mainland. After approximately 10 000 BP, discard of all categories of cultural materials increases with a significant marine component (for the first time) and this peaks during the period 7 800–7 000 BP as the sea approached its current position. When other dated sequences from North West Cape, and nearby Dampier Archipelago and Pilbara coastline are considered (cf. Przywolnik 2005), it is clear that systematic exploitation of marine resources has had an extremely lengthy and near-continuous history on this arid coastal plain.

The mammalian fauna in the Montebello sites are highly diverse and significantly richer than those currently recorded today on nearby Barrow Island. The diversity of medium-sized mammals is also higher than in the contemporaneous Cape Range faunas, primarily owing to the presence of a number of sand plain species assumed to have inhabited the now-submerged coastal plains. It appears that the coastal plain went through a phase of positive water balance, allowing herbaceous plant growth and, of special note, a short-lived south-westerly range expansion of the northern nail-tail wallaby, *Onychogalea unguifera*, a species today found in the Kimberley region of Western Australia.

Littoral resources are believed to have become proximal (i.e. daily walking distance) to the sites by approximately 10 000 BP. A marked increase in the exploitation of marine resources is registered at Noala Cave by approximately 10 000 BP, when shell starts to accumulate in relative abundance. The dense midden of Hayne's Cave, dated to between approximately 7 800–7 000 BP, accumulates at precisely the time when the coast is estimated to be adjacent to the sites. The final phase of occupation of Hayne's Cave is characterised by the loss of most terrestrial fauna coupled with the continued use of marine resources. Species from all marine habitats appear to have been exploited including mature reef flats, rocky foreshore substrates, intertidal mudflats and, importantly, mangrove communities. There are mangrove shellfish, crab and saltwater crocodile recorded from approximately 12 000–7 000 BP. There is little reason to believe that the littoral environment was depauperate either during a transgressive phase or during the period when sea level stabilised at the current high stand.

The final phase of drowning and isolation of the sites from the mainland by circa 7 000 BP is marked by human abandonment, which continued through to the historic period.

The survey and excavation data from the Montebello Islands also allows us to reflect on the unique characteristics of foragers on an arid coastline. The terrestrial faunas of this once extensive coastal plain were more abundant than previously thought. Combined with the accessibility of rich sub-tropical marine species (high in protein and fat) the coastal/sub-coastal zone would have been attractive to hunter-gatherers during the terminal Pleistocene/early Holocene. The usual attributes associated with desert hunter-gatherers (the 'desert adaptation' after Gould 1977) characterised by high residential mobility may well have been transformed in this apparently productive coastal setting.

Veth, Aplin, Wallis, Manne, Pulsford, White and Chappell

Table of Contents

Introduction

The Montebello Islands lie approximately 80 km north-west from the present Pilbara coastline of northwestern Australia (between latitudes 20°21′S and 22°32′S and longitudes 115°31′E and 115°36′E) near the outer margin of the broad, shallow continental shelf (Figures 1 and 2). The islands' small size, degree of remoteness and generally denuded appearance belie the enormous significance they have to Australian archaeology for their ability to inform models about coastal Aboriginal occupation during the terminal Pleistocene and early Holocene (Veth, 1993). They also have the potential to provide critical information on changing terrestrial and marine landscapes during the last two, and possibly three phases of sea level transgression. However, despite their recognised potential, until recently archaeological research on the islands had been minimal, as had interest on the adjacent mainland coast (compare papers in Bowdler, 1984, with those in Hall and McNiven, 1999a).

This situation notwithstanding, north-western Australia in general has attracted considerable interest in relation to regional geomorphology and palaeoclimatology (Webster and Streten, 1978; Semeniuk, 1993, 1996; Kershaw and Nanson, 1993; Wyrwoll, Kendrick and Long, 1993), palynology (van der Kaars, 1991; van der Kaars, Wang, Kershaw, Guichard and Setiabudi, 2000; van der Kaars and De Deckker, 2002), local and regional sea level history (J. Chappell, 1994; Lambeck and Chappell, 2001; Yokoyama, Lambeck, De Deckker, Johnston and Fifield, 2000) and palaeoceanography (Martinez, De Deckker and Barrows, 1999). There is also a rapidly growing body of information on palaeofaunal change in the region, derived primarily from palaeontological sites on Barrow Island and North West Cape (e.g. Baynes and Jones, 1993; Aplin, Baynes, Chappell and Pillans, 2001) and from regional biogeographic studies (Baynes and Jones, 1993; Aplin, Adams and Cowans, in press).
When considered in combination, the various data make it possible to construct a gross framework for landscape evolution and resource availability in the vicinity of the Montebello Islands. As such, the islands provide a unique opportunity to address a number of critical questions related to the history of coastal utilisation in Australian archaeology (e.g. Bowdler, 1984, 1999; O'Connor and Sullivan, 1994; Veth, 1995, 1999; Morse, 1996, 1999; Hall and McNiven, 1999a; O'Connor, 1999a), and to provide additional insights into

various unresolved regional questions in the fields of vertebrate taxonomy and biogeography (Kendrick 1993; Aplin *et al.*, in press).

This monograph makes available for the first time, a summary of the survey, excavation and analysis results from the Australian Research Council (ARC) and AIATSIS funded Montebello Archaeological Research Project (MARP), including detailed information about the extensive faunal remains recovered from two of the excavated archaeological assemblages. It is presented in two sections. The first part of the monograph is devoted to detailing results from the Noala and Hayne's Cave sites, both situated on a peninsula extending from the eastern side of Campbell Island. In the latter section, we turn our attention to considering the Montebello results in terms of the regional archaeological sequence. Throughout the monograph, we draw attention to the following important issues and related questions which shaped the design and execution of MARP:

a) Given their relatively steep shelf profile to the north, the islands provide an opportunity to explore issues of long-term human use of marine systems during the Pleistocene by sampling three proximal shorelines at approximately 40 ka, 30 ka and < 10 ka (cf. Barker, 1999, 2004; Morse, 1999; O'Connor, 1999a, 1999b).

b) Since the Montebello Islands are comprised of Quaternary calcarenite with no known outcrops of volcanic, metamorphic or siliceous materials, any occurrence of such materials amongst the archaeological artefact assemblages would provide excellent evidence for contact and/or mobility strategies between the Pleistocene coastline and the interior, at distances of up to 120 km;

c) The geographical positioning of the islands means they are well situated to potentially record human responses to the last major episode of sea level transgressions. This is especially the case during the period from approximately 10 000–7 000 BP when the shoreline rose to within several kilometres of the islands. A key issue is whether there is a time lag between sea level rise and midden development or whether the appearance of middens is archaeologically 'instantaneous' (after Beaton, 1995);

d) The previous question raises the issue of whether the transgressive shorelines were depauperate in resources or instead provided most of the habitats that were systematically exploited

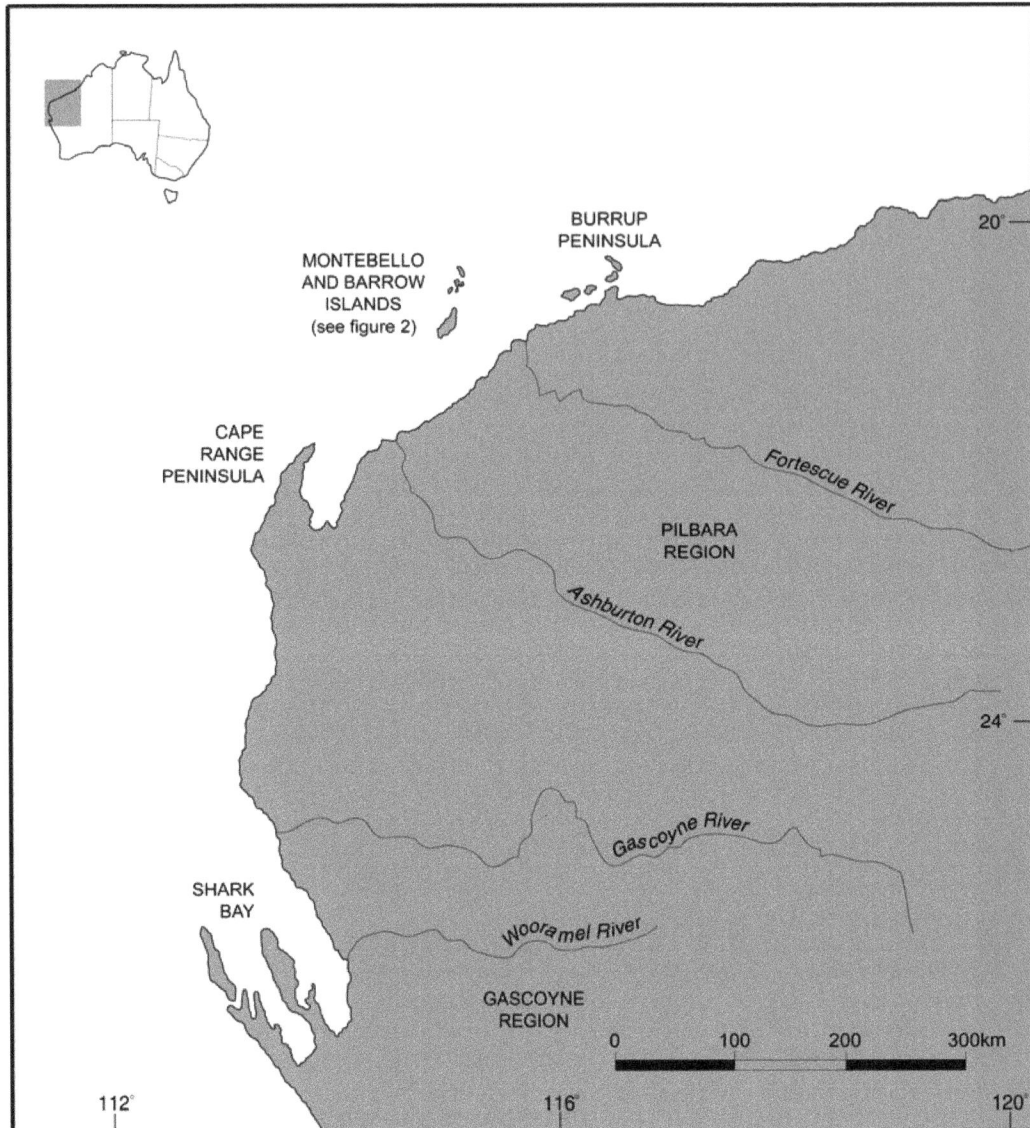

Figure 1   *Map of north-west coast of Australia showing location of Montebello Islands and other key sites mentioned in text.*

by coastal foragers in the recent past. Bowdler (1999:82) specifically posed the question as to when the mangrove/estuarine adaptation, witnessed during the mid- to late Holocene along the central and northern coastline of Western Australia, came into being. Even more fundamentally, we can ask whether mangrove communities were at all present on the edge of the continental shelf at times of lowered sea levels? And, if there were, how rapidly did they re-establish themselves in the context of a coastline moving in response to sea level change?

e) Evidence for close links between environmental changes and human responses can potentially contribute to the emerging debate over the role of environmental factors *versus* social intensification as a prime mover behind the exploitation of maritime resources *per se* (see also Hall and McNiven, 1999b:3; Veth, O'Connor and Wallis, 2000);

f) One question that naturally emerges from any study of remote offshore islands is whether people remained on the islands after marine transgressions, and if so, for how long did the population persist (*cf.* Sim, 1998, 2002, 2004; Barker, 2004)? Similarly, if the islands did not support a resident population following insulation, did mainland or possible Barrow Island residents visit them using watercraft? Given the relatively small size of the Montebello Islands and their close proximity to the larger Barrow Island[1], questions of this nature are perhaps more sensibly framed in terms of the combined Montebello and

---

[1] As discussed later, restricted access to Barrow Island has prevented it having been included in the Research Program.

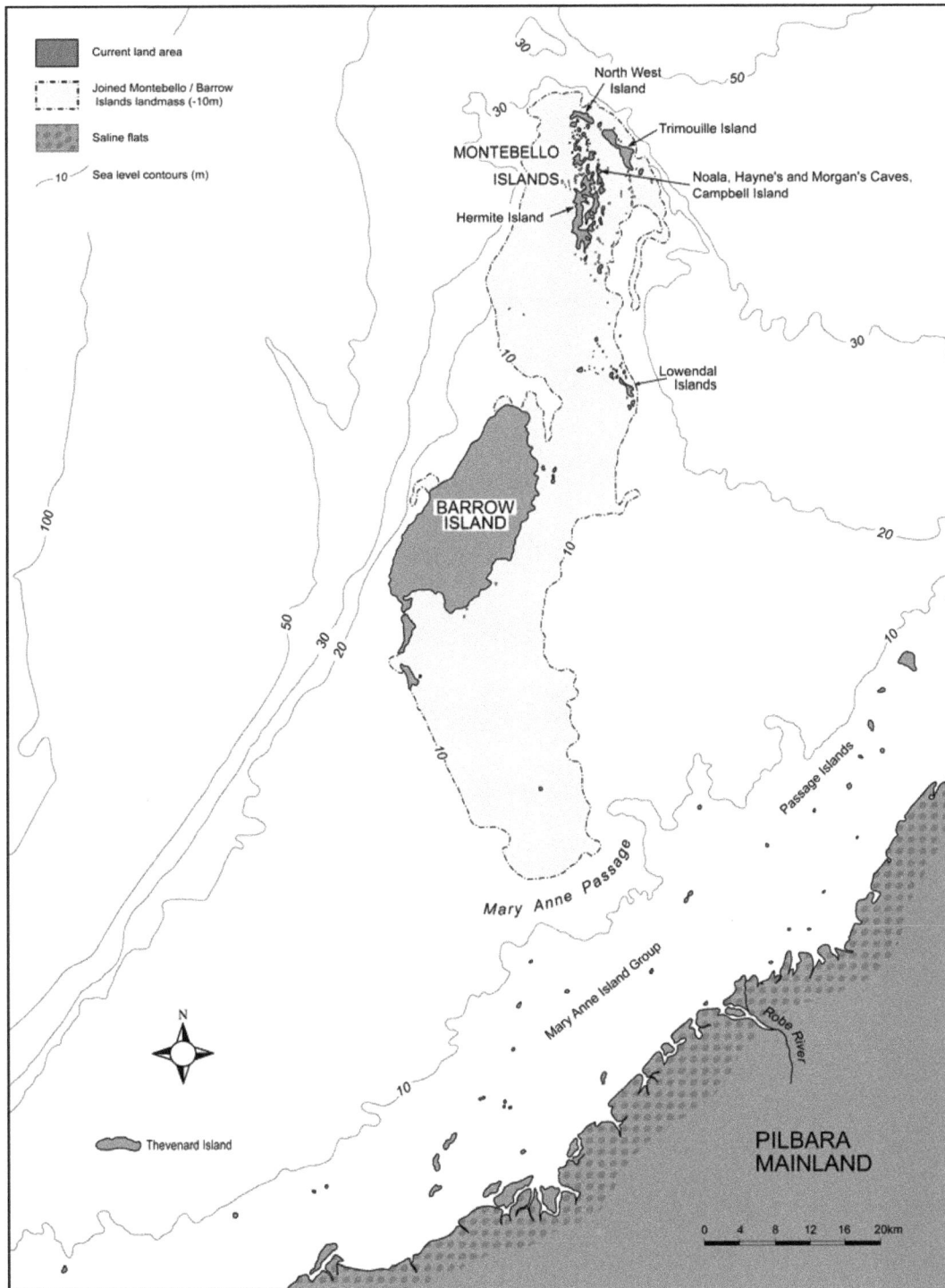

Figure 2   Detailed map of Montebello Islands showing bathymetric contours and joined Montebello/Barrow landmass (- 10m).

Barrow Island group. However, the Montebello Islands would be expected to witness at least occasional visitation from any persistent, residential population on Barrow Island[2].

g) Finally, but certainly not least, the arid coastline of Western Australia is one of the longest on any

continent and would likely have been even more extensive during the Last Glacial Maximum (LGM); literally, it is where the desert meets the sea (*cf.* O'Connor and Veth, 1993). The characteristics of Pleistocene and early Holocene foragers on such an arid coastline and the manner in which they would have structured their use of littoral and hinterland resources is a matter of international interest. The Montebello Islands provide a rare opportunity to examine some of these processes.

Overview of the study area

The present day Montebello Island group mainly comprises small rocky outcrops separated by shallow, narrow channels. The approximately 100 land masses are part of a remnant Quaternary calcarenite plateau which overlies the same Miocene anticline (Tulki Limestone) that forms the core of nearby Barrow Island (Hocking, Moors and van der Graff, 1987). The soft calcarenite is ferruginous, with a resulting reddish colour and exhibits cross bedding, with a strong susceptibility to chemical and mechanical weathering (Smith, 1965). The calcarenite plateau was gradually eroded through the later Quaternary period to form a complex of buttes with cliffy margins that sometimes contain rockshelters – in most cases, however, shelters have collapsed during undercutting, a process still in process today. Following rising sea levels associated with the last marine transgression, the heavily dissected coastal plateau was transformed into the myriad islands of the present Montebello group. While some islands are quite large in area—such as Trimouille at 492 ha—the majority are far smaller (Berry, 1993; also see Figure 2). The channel that separates Trimouille from Campbell Island, where the key archaeological sites are located (see below), is the deepest in the archipelago at approximately 10 m (although this depth is difficult to ascertain with great accuracy owing to the highly mobile marine sediments).

The highly weathered calcarenite of the Montebello Islands typically yields a jagged and often barren land surface. The larger islands have varying degrees of build up of Holocene sand derived from decomposed surficial calcarenite and the reworking of near-shore sand deposits (Burbidge, 1971). The skeletal soils, coupled with the arid climate (mean annual rainfall on nearby Barrow Island is <320 mm; Bureau of Meteorology, 1998) have important repercussions for the floral and faunal suites able to survive on the islands. Temperatures range between 24°C and 35°C throughout the year, with an average humidity of 65 %.

In addition to the low annual rainfall, individual rain events can be sudden and extensive, typically being associated with cyclonic weather activity. Rainfall can result in surficial pooling of freshwater, although such pools quickly disappear and do not offer animals or people a reliable or plentiful resource. The only other source of freshwater during the dry season derives from 'the heavy dew which occurs at night' (Hill, 1955:114). Both the nearby Cape Range Peninsula (on the adjacent Pilbara mainland) and Barrow Island contain perched freshwater reserves known as *Ghyben-Herzberg* lenses (Humphreys, 1993, Cavalche and Pulido-Bosch, 1994; Wicks and Herman, 1995) that feed surface freshwater soaks and springs, many of which are rejuvenated by the cyclonic downpours; however, such lenses are not present on the Montebellos owing to their lower mean height above sea level. There is an historic well associated with an early pearling establishment on the islands (see below), however it is thought that this feature would not have been present during the prehistoric period owing to shifts in the water table and sea level.

It is worth reiterating here that the islands are situated in the middle of the cyclone belt with 16 cyclones having been recorded as passing within 1 km of them over a 15 year period (Berry, 1993), a fact which may have important implications for the survival and integrity of archaeological deposits (cf. Bird, 1992). According to studies carried out on the Burrup Peninsula (located on the nearby Australian mainland) the limit of cyclone effects has been estimated to be 11 m above the lowest astronomical tide (Chappell, 1982). Widespread removal and redistribution of archaeological materials from open midden sites on the nearby Cape Range Peninsula was comprehensively documented by Przywolnik (2002a) following Tropical Cyclone Vance in 1999. She concluded 'The impact of extreme seasonal weather on archaeological sites located on high-energy, exposed coastlines in northern Australia is the most likely explanation for the lack of mounded midden sites in such areas' (Przywolnik 2002a:151), a comment that may be of direct relevant to the Montebello Islands. However, it is also possible that storm deposits of shell materials might accumulate following some of these major storms (cf. O'Connor and Sullivan, 1994); this theme is developed below. The effects of cyclonic and wet season activity on the productivity of shell beds also needs to be taken into consideration in terms of potential marine resource availability (cf. Meehan, 1982).

Between 1902 and 1939 the State Government issued several leases for pearling and the harvesting of dugong and turtle in the Montebello Islands area (Morris, 1991:4). There are a number of wrecks in the waters of the islands that relate to these activities, as well as historic sites associated with T.H. Haynes who took up one of the early pearling leases (Crawford, 1986:6; Morris, 1991:4).

Another part of the island's history that bears consideration is that during the 1950s the Montebello Islands were the scene of the first nuclear tests by the British (Morris, 1991:4), with little recording of current environmental conditions prior to the detonations. Nevertheless, one member of the 1952 Atomic Testing Expedition (Hill, 1955) did undertake some floral and faunal assessments which allow some assessment of the impact of the atomic testing on the local environs. Comparison of Hill's (1955) surveys with the flora and fauna observed by MARP team members, coupled with data from other limited surveys during the 1990s suggest that while the detonations must have had considerable impact in the immediate and short-term, the local flora and fauna appear, at least at a gross level, to have re-established in broadly similar patterns to those pre-blast, for example, decimated mangrove stands along the islands margins are now well re-established. By contemporary standards the detonations were small and the fall out was largely restricted to Trimouille and Alpha Islands. Contemporary reports suggested that much of the aerial fallout drifted onto the Pilbara mainland, a fact which may help explain the similarity in pre- and post-blast plant and animal populations. Prior to and during the first MARP surveys, Government radiation specialists undertook testing and determined that none of the Montebello Islands (away from grounds zero) had elevated radiation readings and were therefore safe for human visitation.

The historical mammal fauna on the islands was extremely restricted. Hill (1955:115) reported finding only a single species of bat (*Eptesicus pumilus*) but also noted that populations of the spectacled hare wallaby (*Lagorchestes conspicillatus*) and the golden bandicoot (*Isoodon auratus barrowensis*) had become extinct since 1912 as a result of the introduction of feral cats to the islands. In contrast, reptiles, especially lizards, were noted by Hill as being numerous, as were birds.

The diversity of plant species on the islands is extremely low and varies depending on the specific environ, which can be broadly classified as (1) rocky crevices, (2) sandy beaches and dunes, (3) inland areas and (4) sheltered sandy embayments (Figure 3). The rocky limestone crevices support a small number of herbs and low shrubby plants, such as *Brewia media*, *Euphorbia* spp., *Phyllanthus* sp., *Polanisia viscosa*, *Capparis nummularia*, *Melhania incana*, *Oldenlandia* sp. and *Nicotania* sp. (Hill, 1955:116). Beach and dune areas are dominated by *Lepidium pedicellosum*, *Frankenia pauciflorum*, *Sesuvium portulacastrum*, *Pimelea ammocharis*, *Sarcostemma australe*, *Calandrinia* sp., *Ipomea pes-caprae* and *Spinifex longifolius* (Hill, 1955:117). *Triodia wiseana*, *Pterigeron decurrens*, *Pterocaulon sphaeranthoides*, *Scaevola* spp., *Canavalia rosea*, *Swainsona* spp., *Indigofera* spp. and acacia (*Acacia coriacea*) dominate the raised inland areas (Hill, 1955; Beard, 1975). The only trees on the islands are pockets of mangroves (*Avicennia*, *Rhizophora*, *Bruguiera* and *Ceriops*) which persist in the sheltered embayments (Hill, 1955; Johnstone, 1990). Although some dense spinifex stands are found in the interior of the larger islands, overall the level of vegetation cover is minimal, resulting in extremely high ground surface visibility (>95 %).

At present the Montebello and Barrow Islands are part of a marine conservation reserve, established on 10 December 2004 (CALM, 2005). The reserves are entrusted to the Marine Parks and Reserves Authority and managed by the Department of Conservation and Land Management. There is no permanent habitation today on the islands, although a pearling company maintains a semi-permanent presence on yachts, houseboats and work-sheds.

Tindale (1974: 254) and Horton (1994:803-4) described both the Barrow and Montebello Islands as lying within the territory of the *Noala* tribe, whose territories encompassed the:

> Coastal plain from about Cape Preston near the mouth of Fortescue River southwest in a strip about 40 miles (65 km) wide to a line running south from Onslow, but not extending to the Ashburton River, which is held by the Talandji. They kept near the seashore and went out to the Barrow and Monte Bello islands using a form of wooden canoe.

However, as Veth (1993) noted, there were no known living Noala speakers with knowledge about human use of the Montebello Islands, nor any mythological tracks documented in the Pilbara that include them.

Previous archaeological research in the region

Barrow Island, a substantially larger continental shelf island, lies between the Montebello Islands and the current Pilbara coastline and has been the primary focus of limited off-shore archaeological research in the immediate region. While stone artefact scatters, shelters with evidence for prehistoric occupation and historic sites with evidence for Aboriginal contact in the form of flaked glass have been noted as occurring on Barrow (Dortch and Morse, 1984; Hook, McDonald, Paterson, Souter and Veitch, 2004; Al Paterson, pers. comm.), to date no site descriptions have been published. Nevertheless, the mere existence of sites on Barrow Island has led to considerable speculation that Aboriginal groups may have travelled there in recent prehistoric or even historic times from the mainland and, if this were possible, further on to the Montebello Islands.

A preliminary investigation of the Montebello Islands by Crawford (1986) failed to locate any

Figure 3   Photograph of typical calcarenite rocky crevice environment and sandy embayment.

Figure 4   View of partially collapsed entrances to Haynes's Cave (LHS) and Noala Cave (RHS), (Campbell Island).

Aboriginal sites, although he did note the presence of historic debris associated with pearling, turtle farming, aquaculture and the 1950s nuclear testing program. By way of partial explanation for the absence of finds, Crawford (1986: 2) argued that the 60 km trip from the mainland to Barrow Island and then onto the Montebellos would generally have been beyond the limits of raft users (although such an explanation could not account for the absence of sites pre-dating insulation).

Montebello Archaeological Research Program (MARP)

Systematic research into the Aboriginal archaeology of the Montebello Islands began with an AIATSIS research grant to Peter Veth in 1991, followed by an institutional (University of Western Australia) ARC grant to Veth in 1992 and

continued under a final institutional (James Cook University) ARC grant in 1994.

During September 1992 a three week field program of survey, mapping and excavation was carried out by a team of six researchers and volunteers. This field season was followed by a second season in October 1994 by a team of six members. Primary transportation to and from the islands was provided by a chartered commercial fishing vessel, whilst transportation between islands for the duration of fieldwork was via a small aluminium runabout.

The extremely rich assemblages recovered during excavation were returned to the mainland at the end of each field season and laboratory analyses continued until immediately preceding the publication of this monograph. Materials are currently stored at the Western Australian

Museum and the Australian Institute of Aboriginal and Torres Strait Islander Studies.

Archaeological survey results

Most of the larger islands in the Montebello group that could potentially have supported human habitation were visited by the research team during the initial 1992 field season. Typically these islands were subject to systematic pedestrian survey of their margins, comprising parallel traverses following the outline of the coast and spaced at approximately 20 m distance from each other. In addition, linear traverses were made at approximately 100 m intervals into the interior of the larger islands including Ah Chong, Alpha, Bluebell, Campbell, Delta, Hermite, Northwest, Primrose, South East and Trimouille. Alpha and Trimouille were surveyed by Veth whilst in the company of the Health Department's Radiation Officers.

A complex of comparatively intact small caves and interconnected chambers was discovered at only one locality within the entire island group; at the end of a small, low limestone peninsula extending east from Campbell Island (Figure 4). Only one small cave on the northern side was visible from the waterline (Noala Cave) while on the southern flank small openings and crevices provided entrance into a partially collapsed system of chambers, named Hayne's and Morgan's Caves. On initial inspection all caves had surface evidence for previous occupation in the form of diagnostically mainland (volcanic/metamorphic) stone artefacts, dietary fauna and organic staining. The smallest of these, Morgan's Cave, produced a small quantity of material that remains unanalysed other than for a cursory examination of the small quantity of faunal remains by Aplin. The Morgan's Cave results will be discussed in a future publication.

With the exception of the shelters, the only surface finds clearly associated with Aboriginal occupation of the islands were three flakes recovered from the central group of islands. These artefacts were all heavily weathered, showed no evidence for retouch or utilisation and were produced on a dense, weakly silicified calcrete of a kind which is widely available on Barrow Island (Ken Aplin, pers. obs.; see below). No outcrops of this material were located on any of the surveyed Montebello Islands.

A linear stone arrangement was recorded in the intertidal zone within the central islands; however, this feature is believed to date to the time when Malays were turtle farming in this area. It is located near some stone weirs and hearth bases that are known to date from this historic activity (Crawford, 1986; Morris, 1991).

Of greater interest were a number of mounded shell features located just above the supratidal zone on several of the central islands and superficially taking the appearance of middens. The mounds are concentrated in small depressions and appear to have been oriented through water action. Two 25 x 25 cm test-pits were excavated into these features and they were found to be comprised entirely of the marine gastropod *Terebralia* sp. and coral grit. The gastropods also exhibited a high degree of uniformity in size and, given the associated coral grit, this strongly suggests a natural origin (cf. O'Connor and Sullivan, 1994). Furthermore, the mounds contained no artefacts or manuports, charcoal or any other indicators of dietary behaviour (such as fish otoliths). The mounds and their surrounding sparse shell scatters are believed to represent storm deposits from reworked shellfish beds that are naturally found across the mudflats fringing the mangrove stands. Similar shell deposits have been noted from the supratidal zone adjacent mangroves on the Pilbara coastline (Veth and O'Brien, 1986).

Noala Cave

Context

Noala Cave is a small north-facing chamber with large roof-fall fragments partially blocking its entrance (Figure 5). The cave is situated approximately 5–7 m above the high tide mark and 30 m back from the margin of an extensive calcarenite rocky platform. It is approximately 5 m deep, varies from 1–2 m in height, and only provides partial shelter from direct sunlight and the currently prevailing winds. The roof-fall has clearly entrapped deposits characterised on the surface by reddish-yellow sands containing burnt *Terebralia* sp., *Melo* sp., *Geloina* sp. and weakly silicified calcrete artefacts.

Excavation details

The exposure of rocky shelves and bedrock towards the rear of the cave indicated that appreciable deposits were more likely to occur towards the front of the shelter near the drip-line. In 1992 a 1 x 1 m test-pit was excavated (Square 1), followed by a 50 x 50 cm pit to the west of this (Square 2; see Figure 5). In 1994 the intervening deposit was excavated (Square 1A) thereby providing a continuous section across the mouth of the chamber. The 1994 excavation of Square 1A was undertaken with high precision, with a view to clarifying issues of stratigraphy and dating raised during the 1992 field season.

The excavation of Square 1 in 1992 and excavation of Square 1A in 1994 proceeded by arbitrary spits of 2–3 cm or by natural units, with volumetric control and *in situ* plotting. Square 2 was more 'exploratory' and taken in larger 4-6 cm spits (on average). All deposit was passed through nested 6 mm and 3 mm sieves and residues were

Veth, Aplin, Wallis, Manne, Pulsford, White and Chappell

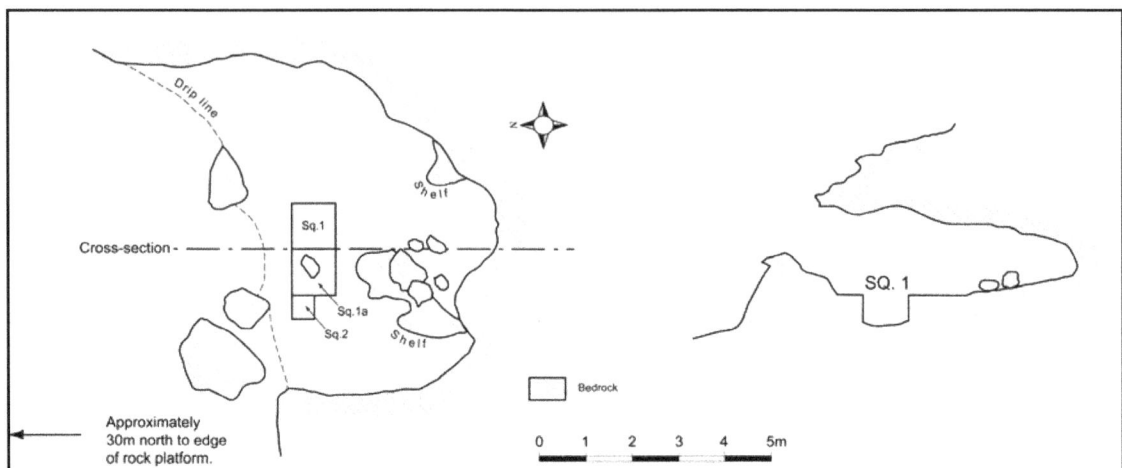

Figure 5   Noala Cave plan of test-pits and cross-section.

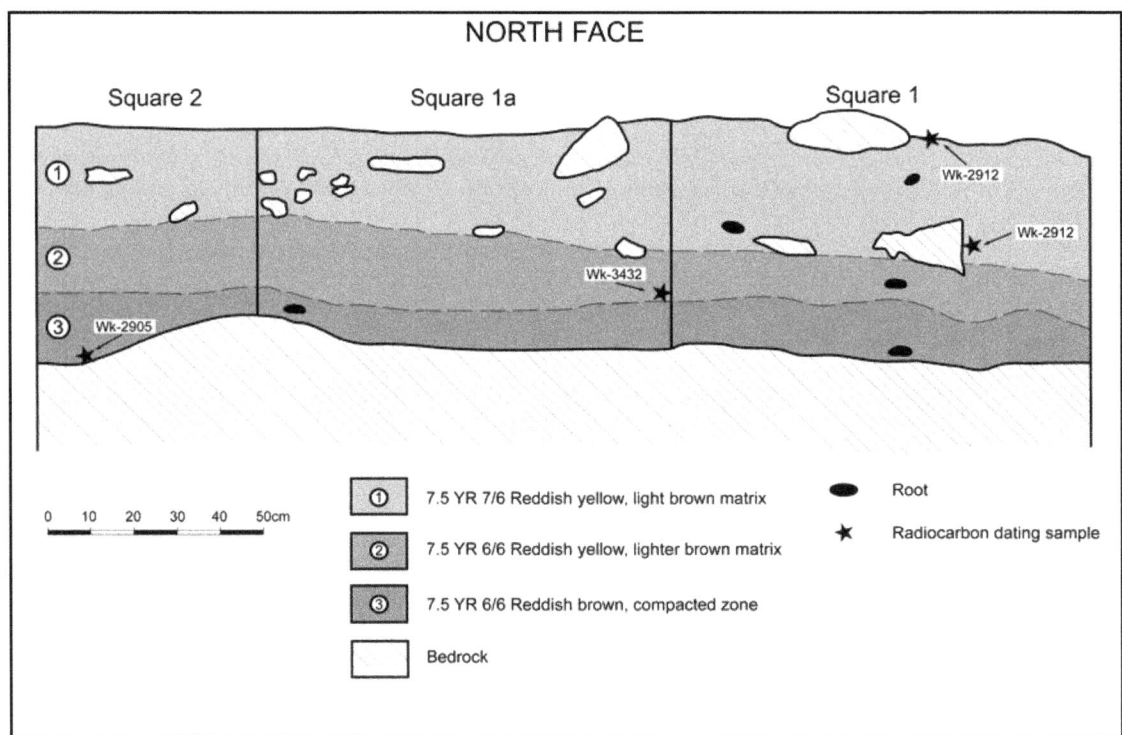

Figure 6   Northern section of Noala Cave, Squares 1, 1A and 2.

bagged and returned to the laboratory for sorting, identification and analysis. Extensive dust amongst sieve residues from Square 1A resulted in the need to wet sieve these samples (using a 1 mm mesh); wet sieving was not required for materials from Squares 1 or 2.

Stratigraphy and sedimentary analyses
When first excavated in 1992, the Noala deposit only had minimally discernable stratigraphy, a feature further clarified during the excavation of 1994. Three minimally differentiated stratigraphic layers (comprising multiple spits) were identified across all test-pits (Figure 6). Discerning different layers at the time of excavation of the three test-

pits (which were joined in 1994) was extremely difficult due to very minor variations in colour and compactness recorded during the removal of each square and then in section. These apparent layers are now not thought to crosscut the more significant changes in fauna, artefacts and sediment uncovered during subsequent laboratory analyses. We would stress that data should be examined by spit and square generally rather than collapsed into these layers (depth below surface data was collected for all spit ends). As noted, average spit depths from Square 2, removed at the end of fieldwork in 1992, were greater than Square 1 (*circa* 5 cm) – and this is reflected in the smaller spit number from which the earliest Baler shell fragment was obtained – in comparison to dates from Squares 1 (1992) and Square 1A (1994) where the spit numbers are higher. In balance the then apparent layers are now not thought to be culturally significant, although the compactness of Layer 3 may indeed be a product of induration – being consistent with an uncalibrated date of 27,220 BP. At the time of excavating, however, layers were (reasonably) thought to be significant and hence dateable samples sought from their boundaries.

Following excavation, mineralogical, chemical and particle size laboratory analyses of bulk-sampled sediments were undertaken. In summary, 10 samples were analysed with a Siemens D5000 X-ray Diffractometer to characterise general and clay mineralogy, an Inductively Coupled Plasma Mass Spectrometer to test for various trace elements, a Laser Particle Size Analyser to determine particle size distributions, and an Electron Probe Microanalyser to investigate particle morphology with Energy Dispersive Spectroscopy employed to chemically characterise other features of interest.

Generally, all analyses revealed only minor changes in mineralogy throughout the deposit, a feature consistent with the minimally differentiated stratigraphy. Particle size distribution varied little across samples. The majority of the well-rounded particles, other than calcite, are clay or quartz with a skin of either secondary calcite or aragonite. The mineralogical analysis showed high levels of magnesium calcite ($CaMgCO_3$) and minor amounts of quartz and kaolinite and smectite clays in all samples. Consistent changes between the proportions of smectite, kaolinite and aragonite through time (Figure 7) indicate a likely change from arid to more temperate conditions, dating to some time before about 8 000 BP. The presence of authigenic aragonite at levels dating to approximately 8 000 BP and 30 000 BP probably reflects proximal sea-stands at these times.

As shown in Figure 5, Noala Cave has only one entrance and, unlike the adjacent Hayne's Cave (see below), has no openings in the roof through which extraneous sediments or bones may have been introduced into the deposits. Given its relative proximity to the sea it might be expected that the cave deposits would include materials deposited during storm events. However, only negligible quantities of foraminifera, small bivalves and coral fragments (two pieces) were recovered from the excavations, strongly suggesting storm wash has not played a significant role in the sedimentary history of the cave.

Dating
The strongly alkaline pH of 9.5–10 registered throughout the Noala deposit has not been conducive to the preservation of charcoal and no features such as hearths, ash lenses or pits were noted during excavation. In contrast, shell and bone preservation in the sites was very good. Consequently, relatively robust shellfish specimens recovered from the deposit were used for dating purposes, providing four radiocarbon dates (Table 1). All shellfish specimens selected for dating were aragonite secretors and were tested for secondary recrystallisation.

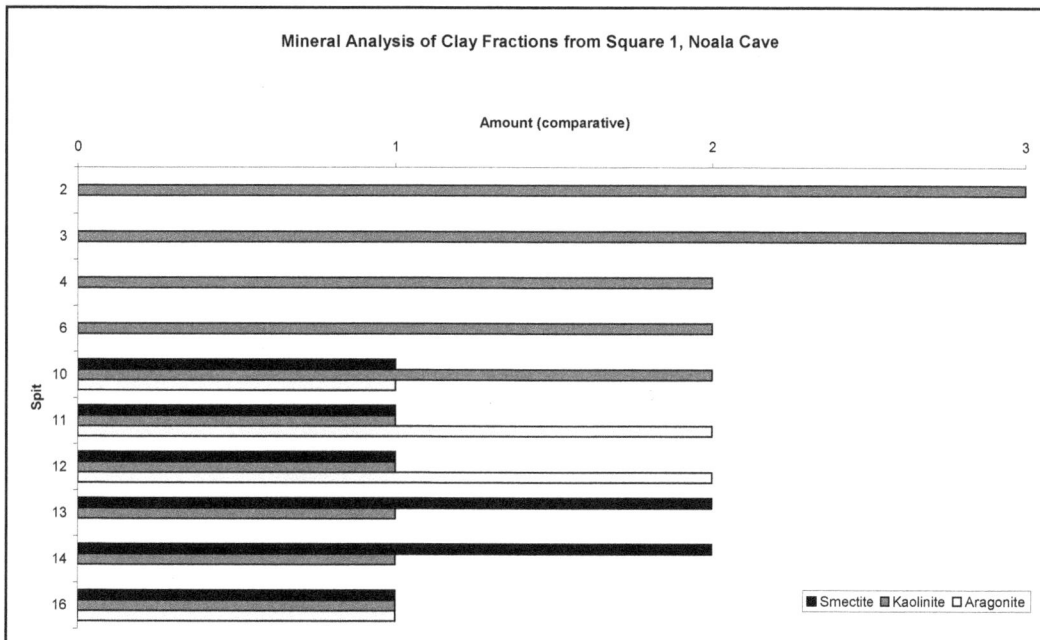

Figure 7   *Graph of mineral analysis of clay fractions from Square 1, Noala Cave.*

Table 1   Summary of radiocarbon dates obtained for cultural deposits on the Montebello Islands.

Site	Lab Number	Square	Spit	Sample Material	Weight (g)	Uncalibrated Age
Hayne's Cave	Wk-2915	2	6	*Terebralia* sp.	12.48	7,930 ± 120
	Wk-2910	3	1	*Polymesoda* sp.	14.11	7,180 ± 110
	Wk-2911	3	7	*Terebralia* sp.	19.87	8,240 ± 90
	Wk-2906	4	1	*Terebralia* sp.	64.88	7,460 ± 70
	Wk-2914	4	3	*Melo* sp.	70.9	8,090 ± 70
	Wk-2907	4	5	Charcoal	4.74	7,630 ± 280
	Wk-2908	4	8	*Terebralia* sp.	53.77	7,890 ± 70
	Wk-2909	4	12	*Terebralia* sp.	52.92	7,820 ± 70
	Wk-2719	4	13	*Terebralia* sp.	25.22	7,560 ± 70
Noala Cave	Wk-2912	1	1	*Terebralia* sp.	21.7	8,730 ± 80
	Wk-2913	1	6	*Terebralia* sp.	6.86	10,030 ± 200
	Wk-3432	1A	12	*Melo* sp.	27.9	12,440 ± 110
	Wk-2905	2	7	*Geloina* sp.	9.2	27,220 ± 640

The samples were collected *in situ* and, in ascending order of likely age, were chosen in order to date the surface and base of Layer 1, the base of Layer 2 and the oldest datable organic materials from compacted (and thought to be older) Layer 3. It should be pointed out that the Pleistocene date on the *Geloina* valve is likely reading up to two sigma too young, exacerbated by the relatively small size of the sample (Alan Hogg, University of Waikato, pers. comm.).

On the basis of the available dates, it appears that Noala Cave was briefly occupied at approximately 30,000 BP, when the islands comprised part of the Australian mainland with the coast having been more than 10 km to the west of the site, based on general sea level curves (cf. Veth 1993; A. Chappell, 1994; see Figures 1 and 2). Only a small volume of sediments and sparse cultural remains separate this date from that of 12,440 BP at which time the sea was approximately 10 km from the site. This apparent hiatus in occupation during the LGM period is a common feature of archaeological sites along the northwest coast (Morse, 1988, 1993b; O'Connor, 1999a; Pryzwolnik, 2002b, 2005). Wk-2913 (boundary between layer 1 and 2) at 10 030 BP dates to when the sea encroached to within < 10 km of the site, while the surface date of *circa* 8 700 BP likely indicates cessation of use of this site. It is worth noting that the earliest dates for occupation of nearby Hayne's Cave immediately post-date the terminal date for Noala Cave.

Stone artefacts
Stone artefacts occurred only in small numbers throughout the Noala deposit, with for example, a total of 48 lithics recovered from Square 1A (Table 2) – of which only 19 are certainly of cultural origin. The artefacts are described in detail in Appendix 2. A low artefact density is typical of midden sites dating to the early Holocene along the Pilbara coast (Clune 2002; Elizabeth Bradshaw, pers. comm.) and hence not entirely unexpected. In terms of artefact types the assemblage comprises flakes, broken flakes and flake fragments, one multi-platform core and one edge-altered flake. Artefacts can be separated into those manufactured from 'local' versus 'exotic' sources. Local sources are defined here as weakly silicified calcrete (calcarenite), a source widely available on Barrow Island where it caps ancient alluvium that occurs in valley systems along the western margin of the island (Ken Aplin, pers. obs.). As noted above, no sources of this material were located on the Montebello Islands themselves during survey. In contrast, exotic sources include silcrete, quartz and volcanic materials which must have been traded from areas beyond the contemporary Pilbara coastline, where appropriate geological formations are found (Hocking *et al.*, 1987). Importantly, artefacts made from exotic stone were recovered from different levels of the Noala Cave deposit (i.e. covering both the Pleistocene and Holocene periods), suggesting continuity in the chain of connection from the site to the mainland. Significantly, this is not the case for Hayne's Cave (see below).

Shellfish and crustacea
Marine fauna were recovered in varying abundance from the upper middle and lower spits of Noala Cave (Tables 3 and 4[4]). They were

---

[4] Nomenclature for the molluscan fauna follows Wells and Bryce (1988), except for the genus *Terebralia* where we follow the taxonomic revision of Houbrick (1991).

Table 2   Stone artefact data, Noala Cave.

Noala Sq 1A spits	Number	%
1	1	2
2	4	8
3	4	8
4	4	8
5	5	10
6	5	10
7	5	10
8	0	0
9	3	6
10	6	12
11	6	12
12	3	6
13	2	4

identified by comparison with reference collections held by the Western Australian Museum, the Queensland Museum and James Cook University. The methodology for the determination of Minimum Numbers of Individuals (MNI) for shellfish followed Bowdler (1983), although these figures were only calculated using the 6 mm fraction remains owing to the extreme fragmentation of the 3 mm fraction. Marine shellfish were categorised as either 'economic' or 'non-economic', with the former being defined as those species recorded ethnographically as having been utilised by Aboriginal people as a food or raw material source and of sufficient size to have warranted their exploitation, and the latter being defined as everything else. MNI values for all marine shellfish and the subset assumed to be economic from Square 1A in Noala Cave are presented in Table 3. Total weights by species are presented in Table 4, with corrections made for volumetric differences between spits. Preliminary assessment indicates a similar range and quantity of marine shellfish are present in the other excavated squares.

A limited range of shellfish species make up the bulk of the marine material, dominated by several species of *Terebralia*, with lesser quantities of *Acanthopleura spinosa*, *Cerithidia redii*, *Melo* sp., *Saccostrea* sp. and *Turbo cinerus*. A detailed survey of the contemporary marine molluscs of the Montebello Islands (Wells, Slack-Smith and Bryce, 1994) detected all of the shellfish identified from Noala Cave.

Specimens of the exclusively mangrove mudflat dwelling genus *Terebralia* that could not be confidently assigned to either *T. palustris* (giant mud-creeper) or *T. semistrata/sulcate* (striate mud-creeper) make up the bulk of the upper midden in Noala Cave, with a further five small fragments of *Terebralia* in Spits 7–11 (total weight = 488.18 g). The *Terebralia* specimens appear to have been processed in a manner consistent with that utilised by the Anbarra people in Arnhem Land as recorded by Meehan (1982); they are typically broken in two, with the base broken off in order to recover the interior flesh. A greater proportion of the *Terebralia* remains are

burnt than unburnt, providing further evidence of their anthropogenic origins.

In northern Australia there are several species of rock oyster that live in aggregated masses forming distinct intertidal 'oyster zones' along rocky shores' (Wilson, 2002:46). Owing to their high level of fragmentation and other associated difficulties in species identification, the Noala Cave specimens have only been identified to genus level (*Saccostrea* sp.). They comprise a small component of the shell assemblage the upper spits, with a total weight from Squares 1 and 1A of 85.15 g.

Cerithida redii was found consistently in the upper spits, with a peak in Spits 4 and 5. This is a mud-creeper typically associated with mangroves.

Melo sp., or baler shell, is common to the intertidal and sub-tidal sand zones and muddy reef substrates (Coleman, 1975). It was commonly used throughout northern Australia for a range of purposes including as pendents, spear-thrower discs, hafted scrapers and adzes, spoons and containers (often for water storage; Akerman, 1973; Scall, 1985), and occurs in Noala Cave as small fragments in Spits 1, 2, 4 and 5. A baler fragment was also recovered from the basal spit of Square 2, the dating of which returned an age of 27 200 BP. Baler shell is known to have been a long distance trade item throughout both the Holocene (Smith and Veth, 2004) and late Pleistocene (O'Connor, 1999a).

Acanthopleura spinosa (spiny chiton) is a member of the chiton family that is found on rocky platforms of the upper intertidal zone, growing to a maximum length of 100 mm. It is present in Square 1A of Noala Cave in small numbers of Spits 4–6. It is possible that these specimens many also belong to *A. gemmata*, the primary difference being that the latter has minute but numerous girdle spines, as opposed to the long girdle spines of *A. spinosa* (Wilson, 2002:28).

Turbo cinerus is commonly located on rocks in the intertidal zone, and occurs in minimal quantities at Noala Cave, being represented by single specimens each in Spits 4 and 5 of Square 1A. Both of these specimens have most of their anterior and middle shell removed with only the spiral remaining, probably indicating their processing to facilitate access to the interior flesh.

It is apparent from both the MNI counts and the corrected species weights that both economic and non-economic shellfish are essentially absent from the lower Pleistocene aged spits (with the exception of five small fragments of *Terebralia* in Square 1A and the dated *Geloina* valve from Square 2) and have only a minor presence in the middle spits. Significant quantities of economic

Table 3   *MNI values of marine shell recovered from Square 1A, Noala Cave.*

Spit	1	2	3	4	5	6	7	8	9	10	11	12	13	Totals
Economic Species														
Acanthopleura spinosa				1	2	1								4
Cerithidia redii	2	1	2	7	11	2								25
Melo sp.	1	1		1	1									4
Saccostrea sp.	3	3	2	2	3	1								14
Terebralia palustris	9	6	5	1	3									24
Terebralia semistrata	1		4	3	2									10
Terebralia sp.		14	9	20	23	5	1	1	1	1	1			76
Turbo cinerus				1	1									2
Total	16	25	22	36	46	9	1	1	1	1	1			159
Volumetrically adjusted total	36	50	48	56	51	14	2	2	1	2	2			264
Non-Economic Species														
Barbatia amygdalumtostum				1										1
Cominella acutinodosa	1				1									2
Fragum sp.	1			1										2
Hemidonax sp.	1													1
Patelloida saccharina	2													2
Pterygia sp.	0	2												2
Turbo coliaceus	1													1
Total	6	2		2	1									11
Volumetrically adjusted total	14	4		3	1									22
Combined total	22	27	22	38	47	9	1	1	1	1	1			170
Volumetrically adjusted total	50	54	48	59	52	14	2	2	1	2	2			286

Table 4   *Weights (g) of marine shell recovered from Square 1A, Noala Cave.*

Spit	1	2	3	4	5	6	7	8	9	10	11	Totals
Economic Species												
Acanthopleura spinosa				0.39	2.57							2.96
Cerithidia redii	1.02	0.38	2.58	4.34	8.38	1.59						18.29
Melo sp.	13.88	0.34		2.66	6.46							23.34
Saccostrea sp.	25.69	18.30	12.35	10.35	16.65	2.17						85.51
Terebralia palustris	34.11	49.75	42.42	11.69	33.14							171.11
Terebralia semistrata	7.67	7.40	15.46	10.43	12.11							53.07
Terebralia sp.	11.45	35.96	40.18	75.08	82.21	18.52	<0.01	0.33	0.15	0.08		263.96
Turbo cinerus				1.98	1.48							3.46
Total	93.82	112.13	112.99	116.92	163	22.28	<0.01	0.33	0.15	0.08	0.00	621.70
Volumetrically adjusted total	212.03	223.14	247.45	182.4	180.93	33.87	0.00	0.8	0.15	0.12	0.00	1080.89
Non-Economic Species												
Barbatia amygdalumtostum				0.18								0.18
Cominella acutinodosa	0.01				0.03							0.04
Fragum sp.	0.01			0.01								0.02
Hemidonax sp.	0.02											0.02
Patelloida saccharina	0.02											0.02
Pterygia sp.		0.01										0.01
Turbo coliaceus	0.19											0.19
Total	0.25	0.01		0.19	0.03							0.48
Volumetrically adjusted total	0.57	0.02		0.23	0.03							0.85
Unidentified fragments	12.48	7.43	4.73	8.39	8.16	1.49	0.50	0.70	0.78		0.18	44.84
Combined total	106.55	119.57	117.72	125.5	171.19	23.77	0.50	1.03	0.93	0.08	0.18	622.18
Volumetrically adjusted total	240.8	237.94	257.81	195.78	190.02	36.13	0.84	2.49	0.93	0.12	0.30	1163.16

shellfish (along with minor quantities of non-economic shell) are restricted to spits which appear to post-date 10 000 BP (Square 1 Spit 6), when the coast was closer to the site. In the Holocene-aged upper spits, economic shellfish from both mangrove, rocky intertidal and sandy environments are present, although the mangrove mud-dwellers are by far the dominant component of the assemblage.

In addition to the marine shellfish, minor quantities of crustacean exoskeleton and fish bone were also recovered. The latter comprised fragments of mandibles, maxillae and vertebrae belonging to specimens 15-20 cm in length; these were not identified further as their small size suggested they may have been prey items of birds rather than human food items. Crustacean remains were only found in Spit 4 of Square 1A and these could not be identified further (total weight = 0.13 g). Their scarcity is a notable feature of the Noala Cave deposit given their potential abundance in the surrounding marine environment, and is thought to be a consequence of the relatively small size of the excavation and taphonomic processes. Note that despite the use of fine meshed sieves during excavation, no otoliths were recovered from Noala Cave, in contrast to the situation in nearby Hayne's Cave (see below). This is argued to be a 'real' phenomenon owing to the Noala's distance from the shoreline at the time of occupation.

Terrestrial snails
Terrestrial snails occur throughout the deposit, with the largest quantities in the upper spits. Terrestrial gastropods are commonly found today on Campbell Island scattered across the surface of rockshelter deposits and in rock crevices, where they are attracted to the greater humidity and protection against desiccation. Several of the taxa found in the wider northwest region (e.g. *Rhagada* spp.) are able to aestivate for long periods and large numbers may be found clustered together in sheltered positions (Ken Aplin, pers. obs.). No attempt has been made to identify or quantify this material as it shows no evidence of economic exploitation. However, one promising line of future inquiry could be to investigate the noticeable differences in the degree of fragmentation of the terrestrial snails in each layer. These may well provide an index of degree of trampling and hence intensity of site occupation. Further, while it has been suggested elsewhere that land snails also hold value as potential indicators of environmental conditions (e.g. Shackley, 1981; David and Stanisic, 1991; Barker, 2004), given that the taxonomy and ecology of many of the Pilbara species are uncertain (Slack-Smith, 1993), their potential in this instance is minimal.

Vertebrate fauna
The conditions for preservation of bone at Noala Cave are generally good, although soluble salts have encrusted bone from the lower spits and in

some cases caused it to fissure (A. Chappell, 1994). It is also possible that recrystallisation of bones (Figure 8) in this arid sandy matrix may have caused accelerated loss of larger elements owing to their larger average pore size (Hedges and Millard, 1995). These factors need to be kept in mind when considering interpretations of the assemblage described herein, particularly as the encrustation made it difficult to identify some small bone elements. In some instances it was possible to remove the crystalline growth by using stiff brushes and gentle pressure (following Baynes, Merrilees and Porter, 1975).

Analysis of the Noala Cave vertebrate remains focused on the remains from Squares 1 and 1A. The smaller quantity of remains from Square 2 was visually checked for consistency with the sequence obtained from Square 1 but the findings were not quantified. For Square 1A, a total of 14 225 faunal fragments were recovered (total weight = 61.04 g), with 1 848 (13%) taxonomically assigned below Class level. This figure represents slightly more than at Puntutjarpa (9.8%; Archer, 1977) and Serpent's Glen (10.5%; O'Connor, Veth and Campbell, 1998), these being shelter sites in the mainland arid zone with similarly excellent preservation of faunal materials. A slightly smaller quantity of remains was recovered from Square 1 (43.98 g), with a lesser quantity again from Square 2. The terrestrial faunal remains were identified by comparison to reference collection materials held by the Western Australian Museum. Taxonomic descriptions and comments are provided in Appendix I for several mammal taxa of special interest.

Reptiles
Reptilian bone was easily separated from mammalian bone based on morphological criteria. Within the reptile remains, lizard and snake cranial elements and vertebrae were also easily distinguished, with it being possible to further classify the lizard material into agamids, varanids and skinks/geckos (*cf.* Withers and O' Shea, 1993). Within each lizard family, remains were assigned a size class (small, medium, large) but were not identified further except in the case of the larger skinks, where *Egernia* spp. could be distinguished from *Tiliqua* spp. on the basis of tooth morphology. Difficulties in establishing genus and species for many of the reptiles made MNI and NISP calculations of no analytical value. The occurrence of the various reptile taxa and their weights are summarised in Table 5.

The snake remains were examined by Dr John Scanlon (formerly James Cook University, now Riversleigh Fossils Centre, Mt Isa) who was able to identify a minimum of five species: a Death Adder (*Acanthopis* sp.), a Whip Snake (*Demansia* sp.), the Mulga Snake (*Pseudechis australis*), the Olive

Table 5 *Weights (g) of reptile bone from Squares 1A and 1, Noala Cave (? = uncertain identification). The following categories may represent the following species: Elapid A = Pseudechis sp.; Elapid C = Demansia sp.; Boid A = Antaresia stimsoni; Elapid B was characterised by chunky vertebrae.*

Spit	1	2	3	4	5	6	7	8	9	10	11	12	13	14	15	16	Total
Square1A																	
Small Scincid	0.03			0.01	0.02	0.01											0.07
Medium Scincid		0.01				0.01											0.02
Large Scincid						0.09						0.02					0.09
Small Agamid					0.02												0.02
Medium Agamid									0.01								0.01
Small Varanid										0.04							0.04
Medium Varanid							0.06	0.10	0.13	0.84							1.13
Scincid													0.01				0.01
Varanid			0.08														0.08
Lizard	0.11	0.19	0.18		0.22	0.06	0.14	0.08	0.15	0.22	0.07	0.34	0.1				1.76
Medium Elapid B				0.05							1.02						1.07
Elapid	0.04	0.03	0.08	0.21	0.26	0.06	0.09	0.09	0.11	0.93	0.13	0.04					1.63
Boid		0.12	0.05	0.05	0.11	0.04	0.80	0.05		0.11	0.02	0.26					1.31
Egernia sp.	0.05				0.06			0.11					0.16				1.15
Antaresia sp.													0.16				0.16
Total weight	0.23	0.35	0.39	0.32	0.69	0.27	1.09	0.43	0.40	2.14	1.30	0.67	0.27				8.55
Volumetrically adjusted weight	0.52	0.70	0.85	0.50	0.77	0.41	1.83	1.04	0.40	3.15	2.18	0.90	1.43				14.68
Square 1																	
Small Scincid	0.19	1.09	0.71	0.26	0.12	0.02		0.05	0.04				0.01				2.49
Small Gekkonid			0.01														0.01
Medium Gekkonid			0.01														0.01
Small Agamid			0.13	0.28	0.01	0.01			0.01								0.44
Medium Agamid	0.01	0.07	0.01				0.02	0.07					0.02			0.01	0.19
Large Agamid			0.01			0.01			0.01								0.03
Small Varanid			0.01														0.01
Medium Varanid			0.04		0.09			0.09	0.04		0.02	0.01				0.02	0.29
Varanid			0.02														0.02
Small Lizard	0.64	11.82	8.30	3.60	1.07	0.43	0.27	0.23	0.2	0.16	0.02	0.08	0.08	0.04	0.01	0.13	27.08
Elapid A		0.13	0.05														0.18
Elapid B		0.16	0.07														0.23
Elapid C			0.01														0.01
Boid A			0.07	0.09		0.01				0.02							0.11
Elapid			0.07														0.07
Egernia sp.		0.04			0.08		0.02	0.08	0.17	0.18	0.09	0.01	0.02	0.03	0.01	0.11	0.60
Tiliqua sp.		0.01		0.09						0.07		0.18	0.18	0.07	0.08	0.03	0.80
Antaresia sp.				0.04								0.02	0.16				0.08
Liasis sp.				0.12		?0.14	?0.15	?0.09	?0.21	?0.03		?0.12					0.06
Acanthophis sp.					0.61			0.04					0.04				0.12
Demansia sp.											0.02			0.02		0.07	0.69
																	0.11
Total weight	0.84	13.33	9.46	4.39	1.98	0.47	0.32	0.56	0.46	0.43	0.13	0.12	0.51	0.16	0.10	0.37	33.63
Volumetrically adjusted weight	0.84	28.26	20.71	5.66	3.86	0.99	0.65	1.32	0.93	1.05	0.42	0.22	0.71	0.35	0.15	0.45	66.57

Table 6   Weights (g) of economic mammal bone from Squares 1A and 1, Noala Cave. Peramelid refers to either of the species Isoodon auratus or Perameles bougainvillei.

Spit	1	2	3	4	5	6	7	8	9	10	11	12	13	14	15	16	Totals
Square 1A																	
Bettongia barrowensis				0.04		0.03											0.07
Bettongia lesueur		0.04		0.40	0.41	0.43	0.21	0.16	0.16	0.38	0.20	0.28	0.07				2.74
Dasyurus geoffreii						0.07						0.04	0.02				0.13
Dasyurus hallucatus					0.15												0.15
Isoodon auratus	0.02	0.05	0.08	0.04	0.62	0.06	0.16	0.01	0.17	0.09	0.09	0.32	0.09				1.80
Lagorchestes conspicillatus											6.48	0.18	0.15				6.81
Lagorchestes hirsutus								0.22	0.04		0.06	1.39					1.71
Macropus robustus					0.05					0.31	2.71	10.07	0.17				13.31
Macrotis lagotis	0.18			0.15	0.14		0.06	0.01		0.17	0.11	0.10					0.91
Perameles bougainvillei					0.08	0.05			0.14		0.02	0.02					0.32
Petrogale sp.				1.22		0.35	0.06		0.07	0.07		0.07					1.84
Pteropus alecto				1.58													1.58
Trichosurus arnhemensis				0.01				0.04	0.08	0.12		0.04					0.29
Peramelid	0.13	0.16	0.23	0.08	0.35	0.41	0.18	0.19	1.06	0.28	0.19	0.45	0.05				3.76
Large macropod	0.03					0.09	0.09			0.15		2.87					3.23
Medium macropod			0.04		0.12	0.22	0.30	0.01	0.76	0.07	6.06	4.65	1.61				13.84
Total weight	0.36	0.25	0.35	3.52	1.92	1.71	1.06	0.64	2.48	1.64	15.92	20.48	2.16				52.49
Volumetrically adjusted weight	0.81	0.50	0.77	5.49	2.13	2.60	1.78	1.55	2.48	2.41	26.75	27.65	11.43				86.35
Square 1																	
Bettongia barrowensis								0.02									0.02
Dasyurus hallucatus				0.02													0.02
Dasyurus geoffreii					0.03	0.07											0.10
Lagorchestes hirsutus			0.09			0.02											0.11
Perameles bougainvillei									0.04	0.02			0.17		0.12	0.07	0.43
Macrotis lagotis		0.20	0.45	0.17	0.13	0.16		0.05						0.04			0.95
Isoodon auratus		0.30	0.13	0.14	0.16	0.16	0.10	0.13	0.13	0.03	0.05	0.10	0.02	0.02		0.08	1.55
Bettongia lesueur		0.16	0.42	0.34	0.19	0.29	0.09	0.15	0.05	0.12	0.04	1.20	0.20			0.21	3.45
Peramelid	0.23	0.17	0.03	0.49			0.44	0.16	0.22		0.02	0.59	0.34	0.07	0.03	0.23	3.27
Large macropod													0.42				0.42
Medium macropod		0.04															0.04
Total weight	0.23	0.87	1.12	1.16	0.51	0.70	0.63	0.51	0.44	0.17	0.11	1.89	1.15	0.13	0.15	0.59	10.36
Volumetrically adjusted weight	0.23	1.84	2.45	1.50	1.00	1.47	1.27	1.20	0.89	0.42	0.36	3.48	1.61	0.28	0.22	0.72	18.94

Python (*Liasis olivaceus*) and a Children's Python (*Antaresia* sp.). Of these taxa, *P. australis* and one species each of *Antaresia* (*A. stimsoni*) and *Demansia* (*D. psammophis*) are currently found on Barrow Island. The Olive Python is found in rocky gorges in the Pilbara, generally associated with permanent springs (Cronin, 2001; Storr, Smith and Johnstone, 2002). Two species of Death Adder occur in the Pilbara region, the ground-dwelling *Acanthopsis pyrrhus* which favours hummock grasslands in sandy habitats as close as Giralia at the base of Exmouth Gulf; and *A. wellsei* in rocky habitats on Cape Range and in the Pilbara uplands (Aplin and Donnellan, 1999).

We argue that the majority of the small lizards and snakes (e.g. *Demansia* sp.) are likely to represent prey brought into the cave by raptors or other potential predators (e.g. *Dasyurus* spp., the native cats or quolls), being mostly undamaged and only rarely burnt, as well as being smaller than the usual size range of prey typically processed by humans. However, the remains of the larger snakes and lizards, such as the medium- to large sized goannas (varanids), the very large, solidly built skinks of the genus *Tiliqua*, the medium-sized skinks of the genus *Egernia* and most of the snakes, are thought to be of anthropogenic origin. This issue of the anthropogenic nature or otherwise of the remains is discussed in more detail below.

Mammals
A total of 31 different species of mammals was identified in the remains from Square 1A, Noala Cave. Thirteen of these are considered to represent 'economic' food items, based on considerations of body size and on the physical state of the remains themselves. The remaining 18 species are likely to represent the remains of disaggregated owl and/or kestrel pellets, perhaps with some material derived from natural deaths in the cave.

The distribution and weights of the 'economic' mammal remains in each of Noala Cave Squares 1 and 1A are shown in Table 6. The total weights for each spit are generally very low, reflecting the high degree of fragmentation of the bone elements. Values for both MNI and NISP for Squares 1 and 1A are given in Tables 7 and 8, respectively.

Rodents - Family Muridae
A detailed analysis of the rodent remains and other small mammals from Noala Cave will be published separately by Aplin and Dr Alex Baynes (Western Australian Museum) as part of their wider palaeontological studies of Barrow Island. Here, we provide some general comments on the ecological and taphonomic significance of each of the recovered taxa.

Rodent remains occur throughout the deposit but are more numerous in the lower spits (Table 9). In general, these remains are unburnt and relatively unfragmented and on these grounds are argued to almost certainly derive in most part from disaggregated owl pellets. Twelve species are represented, only three of which are found today on Barrow Island, but all of which are recorded from sub-fossil accumulations in the nearby

Figure 8   *Photograph of typical condition of recrystalliseed bone from Montebello excavations.*

Table 7   MNI values of economic mammal bone recovered from Squares 1A and 1, Noala Cave.

Spit	1	2	3	4	5	6	7	8	9	10	11	12	13	14	15	16	Total
Square 1A																	
Bettongia lesueur		1		1	2		1	1	1	1	2	2	2				14
Dasyurus geoffreii											1	1	1				3
Dasyurus hallucatus					1	1											2
Isoodon auratus			1	1	1	1	2	1	1		1	2		1			10
Lagorchestes conspicillatus				1					1		1	1					4
Lagorchestes hirsutus				1							1	1					3
Macropus robustus	1			2	2				2	2		1					10
Macrotis lagotis							1			1	1	1					3
Pteropus alecto									1								1
Perameles bougainvillei				1		1	1	1	2		1	1					8
Petrogale sp.				2		1			1	2	1	1					8
Trichosurus arnhemensis				1		1	1	1	1	1							5
Large macropod	1																4
Medium macropod			1		1					1							3
Peramelid	2	1	3		1	1				3	1		1				11
Total	2	1	3	10	7	4	5	4	7	7	8	12	4				71
Volumetrically adjusted total	9	4	11	16	9	9	10	10	7	16	15	16	27				
Square 1																	
Bettongia barrowensis		1	1	1		2	1	1	1	1	1	1	1			1	12
Bettongia lesueur			1					1	1								3
Dasyurus geoffreii																	1
Dasyurus hallucatus				1	1												1
Isoodon auratus		1	1	1		1	2	1	1	1	1	1					10
Lagorchestes hirsutus			1														1
Macrotis lagotis				2				2				2		1			8
Perameles bougainvillei						1		1	1			1				1	3
Peramelid	1											1	1				3
Large macropod													1				1
Medium macropod		1															1
Total	0	2	5	5	1	3	3	5	3	1	2	5	1	1	0	2	39
Volumetrically adjusted total	1	6	11	7	4	6	6	12	6	3	7	9	4	2	0	2	

Table 8   NISP values of economic mammal bone recovered from Squares 1A and 1, Noala Cave.

1	2	3	4	5	6	7	8	9	10	11	12	13	14	15	16	Total
			1		1											2
1			9	6	6	4	3	6	10	6	5	3				59
							1				1	1				3
										1						1
1	2	4		6	4	7	1	4	2	3	9	2				45
	1									1	3	1				6
	1						1	1		2	3					8
1			3	4			1		4	2	3					18
			1						1	3	1	1				7
		2														2
		1		2	2		1	5		1	1					13
		3		2	1		1	2			2					11
		1			1	2	2				1					7
4	4	5	3	9	11	7	8	10	8	5	8	1				83
				2	1				1		2					6
	1			3	3	3	1	6	2	2	10	4				35
3	2	4	22	19	15	14	7	19	21	19	29	8				182
16	12	22	39	34	47	42	39	35	47	44	66	69				
			1					1								2
	4	4	4	4	3	3	3	1	4	1	8				1	40
					1	2										3
	2	2	7	5	4	5	5	7	2	2	3	1	1		3	49
		1				1										2
	2	1	4					3			2	3	1			16
								1	1		1	1		1	1	6
2	3	1	11	5	7	9	6	7		1	12	11	2	1	8	86
											2					2
	1															1
2	12	10	27	15	17	17	18	16	7	4	26	18	4	2	13	208
2	25	22	35	29	36	34	43	32	17	13	48	25	9	3	16	389

Pilbara or in the Cape Range to the south (Baynes, 2000; Baynes and Jones, 1993).

The water rat (*Hydromys chrysogaster*) is represented by a single tooth from the uppermost level of the deposit. This widespread rodent species generally inhabits freshwater lakes and creeks, but it also occurs on various waterless offshore islands (Watts and Aslin, 1981). It is

moderately common along the western coastline of Barrow Island where it enters the sea in pursuit

of crustaceans and fish (Ken Aplin, pers. obs.). Its occurrence in the Noala deposit might be the result of a natural death.

Three species of hopping mice are represented in the deposit: *Notomys alexis*, *N. amplus* and *N. longicaudatus*. All are denizens of dry, arid environments (Watts and Aslin, 1981). *N. longicaudatus* and *N. amplus* are both now extinct but each had an original range that extended across much of the arid zone. The few historical records suggest that *N. amplus* was generally associated with sand habitats (Baynes and Johnson, 1996), whereas *N. longicuadtaus* may have been an occupant of harder substrates.

Zyzomys argurus, the common rock-rat, was found in small quantities near the base of Square 1A. This species inhabits dry, rocky habitats across northern Australia (Watts and Aslin, 1981); it is abundant today on the barren limestone plateau of Barrow Island.

Five species of pseudo-mice are represented, namely *Pseudomys desertor*, *P. hermannsburgensis*, *P. nanus*, *P. fieldi*, and *P. chapmani*. The first two taxa are widely distributed across arid western and central Australia and are usually associated with sandy substrates. The contemporary range of *P. nanus* populations is primarily across northern Australia, in tropical woodland and grassland communities, but the species also occurs on Barrow Island where it occupies *Triodia* hummock grassland. In the recent past it occurred throughout the Pilbara and south on the Yilgarn Plateau to the northern edge of the southwestern forest block, where it was collected historically (Baynes, 2000; Baynes and Jones 1993). *P. fieldi* today survives only on Bernier Island, where it lives in sandy habitat. Formerly, it had an extensive distribution that included the Nullarbor Plain and central Australia where it probably occupied preferentially on "a non-sandy substrate associated with rocky ranges" (Baynes and Johnson, 1996: 175). *P. chapmani* is restricted to rocky talus slopes in the Pilbara; it lives in elaborate tunnel systems with associated 'pebble-mounds' (Start and Kitchener, 1998).

The single specimen of the genus *Leggadina* from Spit 11 of Square 1A could not be identified to species level. Today, *L. lakedownensis* is found in areas of heavy soils and *Triodia* in the Pilbara; it occurs on Thevenaud Island off the northwest Pilbara coast, but not on Barrow Island.

Table 9   MNI and NISP values for rodent bone recovered from Square 1A, Noala Cave.

	Spit	1	2	3	4	5	6	7	8	9	10	11	12	13	Total
MNI	Hydromys chrysogaster	1													1
	Leggadina sp.											1			1
	Notomys alexis	2	3	2	2	12	6	5	9	13	12	9	14	10	99
	Notomys amplus		1	1	1	1	2	2	1	4	5	3	4	3	28
	Notomys longicaudatus				1		1	1	1	2	3	1	3	3	16
	Pseudomys chapmani						2	3			1	1	3		10
	Pseudomys desertor		1	1		1	1		2		1	1	5		13
	Pseudomys fieldi	1								1	1	1	3	1	8
	Pseudomys hermannsburgensis	3	1	1	1	2	4	6	5	9	3	9	12	7	63
	Pseudomys nanus	2	2	1	1	1	2	3	2	4	4	2	6	5	35
	Rattus tunneyi	1	1	2	4	2	2	4	3	7	10	6	7	7	56
	Zyzomys argurus											1		1	2
	Muridae indeterminate	4	4	1	5	9	9	5	4	9	7	9	12	7	85
Total		14	13	9	15	28	29	29	27	49	47	44	69	44	417
NISP	Hydromys chrysogaster	1													1
	Leggadina sp.											1			1
	Notomys alexis	3	10	5	4	27	9	19	14	36	36	28	42	23	256
	Notomys amplus		1	1	2	1	5	5	1	4	7	9	6	7	49
	Notomys longicaudatus				1		1	1	2	3	5	3	6	4	26
	Pseudomys chapmani						3	5			2	3	4		17
	Pseudomys desertor		3	3		2	2		4		2	4	11		31
	Pseudomys fieldi	1								2	1	1	5	1	11
	Pseudomys hermannsburgensis	3	1	2	1	2	10	17	10	17	6	18	24	13	124
	Pseudomys nanus	4	2	1	2	3	6	6	3	11	6	4	15	11	74
	Rattus tunneyi	2	3	5	10	11	7	7	8	25	19	13	18	15	143
	Zyzomys argurus											2		1	3
	Muridae indeterminate	10	9	4	11	14	18	7	9	21	18	19	24	14	178
Total		24	29	21	31	60	61	67	51	119	102	105	155	89	914

Rattus tunneyi is represented throughout the deposit. Historically, this species had an extensive distribution that included the Pilbara and much of the Yilgarn Plateau, south to the edge of the southwestern forest block (Watts and Aslin, 1981; Baynes, 2000). Today, it persists on several islands in the Dampier Archipelago but not on Barrow Island. . Thoroughout its range it probably was associated with soft grassy habitats rather than *Triodia* hummock communities.

Marsupial carnivores - Family Dasyuridae

Two large bodied marsupicarnivores are represented in the deposit: the western quoll (*Dasyurus geoffroii*) and the northern quoll (*D. hallucatus*). Today, *D. geoffroii* is confined to forested areas of southwestern Australia; however, it formerly ranged throughout the arid zone. The smaller bodied *D. hallucatus* favours rocky habitats in areas of open forest, savannah and woodland across northern Australia; it is abundant today in parts of the Pilbara and its remains have been recorded from surficial cave deposits on the Cape Range Peninsula (Baynes and Jones, 1993). Neither species is found on Barrow Island today.

Four species of small dasyurids were identified amongst the Noala faunal remains: kultarr (*Dasycercus* sp.), mulgara (*Antechinomys laniger*), a dunnart (*Sminthopsis youngsoni*) and a false-antechinus *Pseudantechinus* sp. cf. *P. roryi*. All but the latter inhabit arid, sandy habitats (various accounts in Strahan, 1998). The recently described *P. roryi* is associated with rocky habitats through the Pilbara and adjacent deserts (Cooper, Aplin and Adams, 2000); it is one of only two dasyurids found today on Barrow Island (the other being the tiny *Planigale* sp. cf. *P. maculata*).

Bilbies - Family Thylacomyidae

Scarce remains of the greater bilby (*Macrotis lagotis*) were recovered throughout the Noala Cave deposit. Modern populations of the species inhabit shrub lands and tussock grasslands of the northern Pilbara and it has been recorded from recent cave deposits in North West Cape; it is absent from present day Barrow Island.

Bandicoots - Family Peramelidae

Two genera of peramelid bandicoots are represented in the Noala Cave deposit. The western barred bandicoot (*Perameles bougainville*) was found in small quantities through the deposit. Although this species was once widespread across arid and semi-arid southern Australia, it now survives only on Bernier and Dorre Islands in Shark Bay (Short, Richards and Turner, 1998). It is strongly associated with sandy substrates.

At least one, and possibly two, species of *Isoodon* (short-nosed bandicoots) are also represented. The majority of specimens are consistent in size with modern specimens of the golden bandicoot (*I. auratus barrowensis)* from nearby Barrow Island. However, several isolated teeth are larger than any modern Barrow Island specimen and compare better with examples of a larger form allied to *I. auratus* from the Kimberley region (see Appendix I for further details). This second form is not large enough to be referrable to a third taxon present in northern Australia, the northern short-nosed bandicoot (*I. macrourus*). The occurrence of two forms in the same deposit may indicate that *I. auratus* comprises several different species, or it might reflect rapid changes or instability in body size within a single population, perhaps in response to the development of insular conditions. The two forms are not discriminated in the data tables presented herein, pending further taxonomic study.

Possums - Family Phalangeridae
The northern brush-tail possum (*Trichosurus vulpecula arnhemensis*) is present in small quantities in a number of spits in Square 1A. This species prefers open forest and woodland, but will nest in rocky crevices, caves and termite mounds. It is abundant today on Barrow Island where it feeds primarily on the leaves and fruit of *Ficus platypoda*.

Rat Kangaroos - Family Potoroidae
Similar to the situation with *Isoodon*, the archaeological specimens of *Bettongia* can be divided into represent two distinct forms dependent upon tooth size and morphology (see Appendix 1). A smaller-toothed form is indistinguishable from modern specimens of *B. lesueur* from Barrow and Boodie Islands, while a larger-toothed form agrees well with typical *B. lesueur* individuals that survive today on Bernier and Dorre Islands in Shark Bay[5]. The smaller-toothed form is the more abundant of the two in the Noala Cave deposit (the reverse is true in Hayne's Cave; see below).

The presence of both forms of *B. lesueur* in the Montebello Island faunal assemblages suggests that these 'populations' co-existed in the same region in the not-too-distant past, apparently without undergoing complete introgression. Although not in itself conclusive, this evidence raises the possibility that the two populations represent distinct taxa, thus are probably best regarded as 'sibling species'. Based on their current distributions and patterns of habitat use, it seems likely that they occupied distinct habitats, the smaller Barrow Island bettong using the rocky limestone plateau and the typical burrowing bettong—or boodie—occupying the sand plain habitats of the exposed continental shelf (Short and Turner, 1993). With marine transgression, the latter taxon has evidently contracted in range to the southern part of the Western Australian coastline. Further taxonomic work is currently underway on the status of this species group.

Kangaroos and wallabies - Family Macropodidae
All kangaroo and wallaby skeletal remains from the Noala Cave deposit are extremely fragmented, with tooth fragments often being the only parts surviving. Three medium-sized macropodids were identified, namely the spectacled hare wallaby

(*Lagorchestes hirsutus*), the rufous hare wallaby (*L. conspicillatus*) and a rock wallaby (*Petrogale* sp.).

The two *Lagorchestes* species have almost complementary modern distributions, the spectacled hare wallaby occurring across subtropical northern Australia and the rufous hare wallaby occupying the arid centre. Today, it is *L. conspicillatus* that is found on Barrow Island and in the Pilbara ranges (but not on Cape Range; Short and Turner, 1991), while *L. hirsutus* survives on Bernier and Dorre Islands and in the sandy deserts of central Australia (Short and Turner, 1992). On Barrow Island, the spectacled hare wallaby is most commonly seen in the dense stands of *Triodia* in the valley floors and on the large, vegetated claypans at the northern and southern ends of the island (Ken Aplin, pers. obs.).

The *Petrogale* sp. material is too fragmented to allow definite allocation to species; however, it most likely represents the black-footed rock wallaby (*Petrogale lateralis*) that occurs today on neighbouring Barrow Island and at Cape Range (Baynes and Jones, 1993). Though *P. rothschildi* from the Pilbara ranges is somewhat larger-toothed than the Noala specimens, we cannot unequivocally rule out the possibility that some of the remains might represent this species.

Fragmentary remains of a wallaroo (*Macropus robustus*) are also present among the faunal remains. These are comparable in size to the diminutive Barrow Island wallaroo (*M. robustus isabellinus*) that is abundant today in all habitats on Barrow Island (Short and Turner, 1991). Both *Petrogale* species and *M. robustus* commonly shelter in rocky terrain (Turnbridge, 1991) and their remains are often found in the confines of caves and overhangs as a consequence of natural deaths.

Fruitbats - Family Pteropodidae
Two fragments of the black flying fox (*Pteropus alecto)*, were found in Spit 4 of Square 1A and both may derive from the same individual. Modern populations of this species are found in estuaries supporting mangrove forests and along coastal freshwater courses with paperbarks (Hall, 1998; Vardon and Tidemann, 1997). This species does not roost in caves, preferring tree canopies, and is too large to have been taken by a raptor, hence it is interpreted as an economic food item, albeit an opportunistic one.

Ghost Bats - Family Megadermatidae
Remains of the predatory ghost bat (*Macroderma gigas*) were found in various spits in both Squares 1 and 1A. *M. gigas* is the only Australian carnivorous bat and it is an exclusive cave dweller (Douglas, 1967; Armstrong and Ainstee, 2000). Although ghost bats often accumulate substantial

---

[5] Mainland populations are sometimes distinguished as a separate subspecies, *B. lesueur grayi;* the entire group is in need of taxonomic revision.

quantities of bone derived from their faeces at their roost sites, this material is finely comminuted and highly characteristic (Douglas, 1967). The physical condition of the Noala Cave faunal remains is not consistent with this mode of accumulation and it is argued that the ghost bat remains more likely represent prey items of either a large raptor or the human occupants of the cave.

Taphonomic considerations of the fauna from Noala Cave

Agents of accumulation

Based on the physical character of the remains, the Noala Cave vertebrate fauna appears to be derived from four different sources:
a) human activity, manifest throughout the deposit at varying intensity;
b) an owl roost, mainly active during the deposition of the middle to lower deposits;
c) a kestrel roost, active primarily during the deposition of the upper spits; and
d) natural deaths in the cave.

Each of these agents of bone accumulation tends to produce remains with particular diagnostic characteristics and taxonomic composition.

Owl activity is probably responsible for the bulk of the small mammal remains. These are concentrated below Spit 5 in Square 1A (see Table 6) although they continue at lower frequency to the surface. In general, the small mammal remains are lightly fragmented and only rarely burnt. Today, owls do not appear to be resident on any of the Montebello Islands or on neighbouring Barrow Island, although they have been recorded as rare visitors to both (Sedgewick, 1978; Tim Pulsford, pers. obs. for Campbell Island, October 1994). Their apparent reluctance to remain on the larger Barrow Island is certainly not due to an insufficient food supply, but might reflect a scarcity of suitable roost sites away from the relatively unsheltered coastal cliff-lines. The most likely species responsible for the accumulation of small mammalian remains in Noala Cave is the barn owl, *Tyto alba*, a widespread species responsible for large bone accumulations throughout arid Australia (Morton and Baynes, 1998). Pulsford (1994) observed this species sheltering in both Noala and Hayne's Caves during the 1994 field season. In addition, the occurrence of fresh rodent bones including a cranium and jaws of the introduced black rat (*Rattus rattus*) on the surface of Noala Cave further attest to the occasional use of the site as a roost. However, the fact that this introduced species was not otherwise recovered from the site indicates the genuine rarity of recent owl visitation.

The large numbers of small reptile bones recovered from the upper part of the deposit of Noala Cave require a different explanation, as while owls do occasionally take diurnally-active reptiles such as skinks and agamids (Dickman, Daly and Connell, 1991), these usually form only a minor component of their roost accumulations. Two other raptors deserve consideration. Ospreys (*Pandion haliaetus*) occur in large numbers throughout the area today, but their nests are generally placed in exposed positions such as on rock pinnacles, and they rarely occur far from the coastline. Inspection of prey remains around osprey nests shows a predominance of fish remains, with occasional bones from other small vertebrates (especially birds). In contrast, kestrels feed primarily on insects, supplemented with small reptiles and occasional small mammals. They also commonly perch and roost inside shallow rockshelters. In our opinion, the small vertebrate remains from the upper spits of the Noala Cave deposit were most likely derived from regular kestrel activity in the site. This shift from 'mainly owls' to 'mainly kestrels' during the accumulation of the non-human component of the fauna might have had multiple causes: changing proximity of the site to the coastline; the increasing frequency and/or duration of human visitation to the site; possibly even a long-term shift in climatic conditions, although it is not possible to us to determine which these factors alone, or in combination were responsible for the shift.

The humanly-derived component of the faunal assemblage is distinguished on the basis of its much higher level of fragmentation, the occurrence of burning and the characteristics of the prey items themselves. As indicated above, we consider the bulk of the small vertebrate remains to be excluded from this category. To some extent, this interpretation is at odds with Australian ethnographic literature that contains abundant reference to the consumption of small vertebrates by people following a traditional lifestyle, especially in desert environments (summarised by Tonkinson, 1978 and Turnbridge, 1991). However, the same literature also indicates that animals this small are usually cooked and consumed whole (Turnbridge, 1991), sometimes after pulverisation (*cf.* Gould, 1996). The majority of the small vertebrate bones at Noala Cave are relatively intact, hence they are unlikely to have been prey items of people, even in the rare instances where they are burnt. In our views, burning of these small bones is more likely to have occurred accidentally, through their chance incorporation into a fire or location directly beneath a hearth.

This argument is given additional support by the observation that many of the bones from the larger reptile species are fragmented and/or burnt. This includes the remains of *Egernia* sp. and *Tiliqua* sp. among the skinks, and the majority of the varanid and snakes remains; these are all judged to be of

Veth, Aplin, Wallis, Manne, Pulsford, White and Chappell

anthropogenic origin. However, in making this distinction between the smaller and larger reptiles, we do not wish to imply that the prehistoric occupants of Noala Cave did not catch and consume smaller reptiles in addition to the available, larger species. Rather, we are suggesting that the mode of preparation and consumption of the small taxa may well have precluded their becoming incorporated into (or at least, recovered from) an archaeological deposit.

A common issue in Australian faunal research is the difficulty of distinguishing between genuinely anthropogenic accumulations and those that have either been produced or secondarily modified by carnivore activity, typically the dingo (*Canis familiaris*) or the Tasmanian Devil (*Sarcophilus harrisii*). In the case of Noala Cave, the age of the deposit surely rules out dingo - which is not thought to have been present in Australian before 3 500 to 4 000 years ago (Corbett, 1995) - as a potential factor. To date there is no direct evidence of the Tasmanian Devil in the Montebello Islands or elsewhere in northwest Australia. However, it was certainly present in northern Australia as recently as the early Holocene (Mulvaney and Kamminga 1999:260) and is probably recorded in cave art in the Kakadu region (Chaloupka, 1993). Hence, the possibility certainly exists that *Sarcophilus* was present in the Montebello region and played a role in the formation of the Noala Cave assemblage. However, we argue it is unlikely that devils played any significant role in producing or modifying the Montebello Island faunal assemblages owing to the absence of any surficial tooth marks or fraying of fracture edges among the remains (Marshall and Cosgrove, 1990 and references therein). We also note the relatively high incidence of burning of the larger vertebrate remains and the presence of other indisputably anthropogenic materials (e.g. stone artefact and burnt marine shell) in association with the faunal remains.

Cultural formation processes
A number of important characteristics of the overall Noala Cave faunal assemblage may be attributed to specific Aboriginal behaviours relating to both the procurement and processing of prey.

The gastropod *Terebralia* sp. first appears in Spits 7–10 when the coast was probably greater than 30 km distant. These specimens include a number of juvenile *Terebralia* shells that would have contained minimal quantities of edible flesh and which therefore are unlikely to have been deliberately collected as a food source. We posit two possible explanations for their presence: (1) they may have been accidentally included in moist clumps of mud used to transport the larger, more desirable shellfish over a significant distance to an inland campsite (Clay Bryce, pers. comm.); (2)

alternatively, they may have been attached to pieces of mangrove wood transported for use as fuel or as raw material for construction of implements or shelters.

All of the shellfish remains from Noala Cave are highly fragmented. As mentioned earlier, characteristic processing fractures associated with the extraction of flesh were recorded on many of the *Terebralia* sp. and *Turbo cinerus* specimens (*cf*. Meehan, 1982). The proportion of burnt versus unburnt *Terebralia* sp., compared to other species of shellfish, is also consistently higher. Although some of the fragmentation of economic shellfish species, such as *Terebralia* sp. and *Saccostrea* sp., can be attributed to food processing activity, treadage and surficial exposure prior to burial are also likely to be responsible for much of the damage. The relative contributions of these factors can be difficult to determine in shell assemblages (Mowat, 1994).

The anthropogenic component of the Noala Cave vertebrate fauna also exhibits a high degree of fragmentation. For example, of the 63 macropodid teeth recovered, only six are complete. Gould (1996) attributed a similar degree of fragmentation of teeth in the Puntutjarpa site to deliberate reduction of bones for their contained nutritional value. Although this process might have contributed to fragmentation of bone at Noala Cave, we suspect that other factors—such as pre- and post-depositional weathering—might be more significant in this case [Smith (2000) offers a similar explanation for fragmentation at Puntutjarpa]. One reason for adopting this position is that the Noala fauna includes very little associated cranial or post-cranial material of any kind for the medium-sized to large mammals. This may well reflect the relatively slow rate of sedimentation in the cave, with large pieces of bone being subjected to long periods of exposure to sunlight and fluctuations in temperature and moisture content. In contrast, smaller bones are perhaps more likely to become buried in the loose surface sediment and hence gain protection from these destructive forces. A major implication of this high level of destruction of the larger bones is that little can be meaningfully deduced about the actual intensity of exploitation by the prehistoric inhabitants of the different classes of prey items.

Noala Cave – evidence for a terminal Pleistocene coastal economy?

The regional geology and hydrology of the Montebello area and surrounds may have had a considerable influence on the quality of 'shelter' afforded by Noala Cave during its various periods of occupation. As stated earlier, both the Cape Range area and Barrow Island contain *Ghyben-Herzberg* lenses that provide surface springs and soaks. During the Quaternary, it is likely that freshwater reserves rose and fell with changes in glacioeustasy, similarly to what is thought to have occurred in other carbonate areas in the world, such as Guam and areas of Puerto Rico (Frank, Mylroie, Troester, Alexander and Carew, 1998; Mylroie, 2001). During the LGM when ocean levels were low, rainwater was either lost as run-off or was percolated down through the carbonate rock to the deeper water table (Melim, 1996). As sea levels rose after the LGM, faster recharge of the freshwater aquifer occurred as the freshwater table was propelled closer to the ground surface.

Though the Pilbara experiences very low annual rainfall, when rainfall does occur, it is in sudden, heavy quantities (Wyrwoll et al., 1993). This rainfall pattern would recharge the freshwater lenses quickly, leaving excess freshwater on the ground surface as springs and soaks. This process may have made the Noala Cave area more hospitable during the terminal Pleistocene, not only for humans but also for their mammalian prey. However, as sea levels continued to rise, freshwater reserves within the Montebello Islands would have become contaminated by seawater. In contrast, Barrow Island continues to have perched freshwater due to its greater mean height above sea level. A similar situation pertains in the Cape Range, where Morse (1993a, 1993b) has argued that the presence of freshwater sources was a powerful determinant for the choice and use of sites by Aboriginal people.

In the period between approximately 25,000 BP and 12,000 BP, spanning the LGM and the initial millennia of deglaciation, it is likely that the coastal plain dunefields would have been at least partially mobile (Kershaw and Nanson, 1993; Semeniuk, 1996). Water sources in the form of soaks within the dunefield and springs associated with the limestone plateau may have diminished or dried up altogether over this period, and the coastline may have fallen outside of the 'primary catchment area' of the site (i.e. the area from which the primary subsistence remains were derived, for any given time slice through the archaeological record). This situation might be expected to have changed by approximately 10 000 BP when the sea again came within approximately 10 km of the site.

The lower cultural assemblages from Noala Cave dating to approximately 30,000 BP provide direct evidence for the occupation of a portion of the drowned coastal plain of northwest Australia. At this time the Montebello Islands were part of the Australian mainland and this slightly elevated cave would have been separated from the distant coastline in the west by some 35 km of arid sand plain and dune fields. The lower spits contains negligible marine fauna, comprising a single valve of the mangrove bivalve *Geloina coaxans* (Square 2), an unidentified fish vertebra and several fragments of the ubiquitous mangrove species *Terebralia* sp. Despite their limited quantities, the marine remains do indicate that the site's occupants retained contact with the 35 km distant coastline, which apparently supported macrophytic communities. What this site can otherwise reveal about Pleistocene utilisation of marine resources is minimal, given the reasonable assumption that 'bulky foods such as shellfish are not usually carried by hunters more than a few kilometres from their source' (Meehan, 1982:2).

In Square 1 of Noala Cave, the lower deposit only contained sparse economic terrestrial fauna; however, in Square 1A the quantity of these remains actually peaks in the lowermost spits (Table 7). The reason for this contrast between the two adjacent squares is not immediately apparent. One obvious possibility is that it represents some local disturbance of the lower deposit in Square 1A. Alternatively it might reflect genuine spatial variation in the distribution of vertebrate remains within the lower spits, perhaps related to placement of hearths within the shelter. At any rate, the contrasting explanations do not radically impact on the overall interpretation of this sequence. The vertebrate fauna from the Pleistocene-aged spits contains a variety of medium-sized reptiles as well as a suite of medium-sized to large mammals including euro, spectacled hare wallaby, bettongs and two species of bandicoots (*Isoodon auratus* and *Perameles bougainville*). Overall the faunal assemblage indicates a terrestrially-based economy with a small component of marine resources. The suite of small mammals from the Pleistocene-aged spits points to the presence of an essentially arid local environment, characterised by a mosaic of sandy, clay-loam and rocky habitats. This is precisely what we might reasonably expect to find on and around the low limestone plateau, with its surrounding apron of exposed continental shelf habitat.

Following this brief period of earlier occupation in the Pleistocene, the site is abandoned (lacks datable cultural materials) from 27,000 to 12,400 BP with increasing aridity and distance to the shoreline associated with the lead into and establishment of LGM conditions. A similar pattern is seen in the adjacent Cape Range Peninsula sites, with Mandu Mandu Creek being abandoned from 20–5 ka (Morse, 1993b), C99 from 21–8 ka and Jansz from 31–11 ka (Przywolnik, 2002b, 2005). O'Connor, Veth and Barham (1999) have argued that such chronological hiatus events cannot be explained by erosional mechanisms (as proposed by Smith and Sharp, 1993) and that instead cultural explanations such as changes in settlement patterns and regional migration consequent on coastal reconfiguration during periods of sea level change are more plausible.

Like other sites in the region, reoccupation of Noala Cave recommences during the millennia immediately following termination of the LGM, associated with a change to milder climatic conditions and rising sea levels. Spits 6-12 from adjacent Squares 1 and 1A are bracketed by dates of approximately 12 400 and 10 000 BP, during which time the coast was still probably 10–15 km distant. Economic marine shell is present in extremely small quantities in Spits 7–11 in Square 1A (circa 5 g in total). Medium-sized terrestrial mammals and reptiles are well-represented throughout this part of the deposit, with a slight decrease in abundance between the lower and upper spits. The diversity of both medium- to large mammals and small mammals is higher than in lower spits. The overwhelming impression from this terminal Pleistocene-aged assemblage is of an essentially arid environment but one supporting a variety of different habitat types and a comparatively high regional faunal diversity.

Spits 1-5 (Squares 1 and 1A), approximately bracketed by dates of 10 000 BP and 8 700 BP, presents a dramatic contrast with the lower deposit. During this time sea level rose to within 10 m of its current level and Noala Cave was located only 10-15 km from the contemporary coastline. Around 10,000 BP the Montebello Islands were the most northerly part of a large peninsula, and were joined by a land bridge to Barrow Island. Significantly, there is a substantial increase in both the quantity and diversity of marine fauna, with more than 97% of MNI values for economic shellfish in Square 1A recorded from this Layer (Table 3), reflecting the relative proximity of the coast to the site. Although this increase in shellfish remains is clearly linked to the post-glacial marine transgression, there are several reasons to believe that the bulk of the upper deposit accumulated prior to the arrival of the coastline in the immediate vicinity of the site. In addition to arguments that sea level reached its current level in this region around 6 500 BP, the first is the fact that the majority of the shellfish brought to the site were mud-dwelling whelks (Terebralia spp. and Cerithidia redii) and oysters (Saccostrea sp.), all of which can be transported and kept fresh longer than other species (Bird, 1996). The second is the paucity of fish and crab remains in the deposit, perhaps reflecting the poor survival of these groups out of the marine environment and their well-known propensity to 'go off in the midday sun'.

The vertebrate fauna from the upper spits are similar in both abundance and composition to those of the middle deposit. However, there is a slight reduction in diversity of species, with loss of several mammal species including Perameles bougainville. This evidence points to a possible reduction in the variety of habitats within the

'catchment' of the site, or perhaps of the areal representation of certain habitat types, particularly those occupying sandy substrates. By the end of occupation of Noala Cave, circa 8 700 BP, increasing sea level likely breached the Mary Ann Passage resulting in the combined Montebello/Barrow Islands forming a large landmass separated from the mainland by only about 5 km.

The cultural assemblages from Noala Cave demonstrate that during the terminal Pleistocene Aboriginal groups were exploiting both the northwest coast and hinterland plains, engaged in what may be described as an 'arid coastal plains' economy. The emerged coastal plain comprised an extensive and diverse habitat, one that probably provided at least ephemeral water sources and supported a wide range of potential vertebrate prey species including abundant reptiles and a high diversity of both small and medium- to large sized mammals. Recovery of Terebralia sp. and Geloina sp. from various levels clearly attest to the presence of mangrove stands along the coastline, while the remains of limpet (Patelloida sp.), oyster (Saccostrea sp.) and baler (Melo amphora) demonstrate the presence of a range of other substrate types.

Prior to 10 000 BP the inhabitants of Noala Cave appear to have focused their economic efforts on the hunting of mammals and reptiles from the diverse habitats presented by the coastal plain and limestone plateaux. However, the presence of at least some marine fauna in all levels through the deposit indicates a pattern of intermittent visitation to the site by people who either visited the coast directly or were in or contact with other groups resident along the coastline. Either way, some exploitation of coastal resources was clearly being practiced.

After circa 10 000 BP, there was a marked increase in the quantity of marine economic remains being returned to Noala Cave. From consideration of shelf bathymetry, this upturn in maritime activity occurred around the time that the coastline came to within approximately 10 km of the site. Interestingly enough, the shellfish remains from these levels point to a pattern of selective prey utilisation based on ease of transport, with a strong emphasis on the most durable and easily carried food items.

Based on the single surficial radiocarbon date, the site was effectively abandoned by 8 300 BP (ORE corrected) at a time when the sea was probably less than 5 km distant. As we shall report below, this coincides with the first visible occupation of nearby Hayne's Cave with its spectacular evidence for a fully-fledged coastal economy.

Hayne's Cave

Context

Hayne's Cave is located on the opposite side of the narrow isthmus from Noala Cave. It is situated approximately 15 m from the shore and only 3-4 m above the high tide mark. It faces to the south, looking over generally well-protected and shallow waters. The cave roof has partially collapsed with chambers only accessible through crevices and around large dislodged boulders (Figure 9). The site can be divided into the western-most chamber, which was almost entirely sealed, and the remaining chambers to the east.

Excavation details

The sediments of the eastern chamber were mounded to the rear and contained only sparse surface cultural residues except along the western margin where specimens of *Geloina* sp., *Terebralia* sp., *Melo* sp. and silicified calcrete artefacts were noted. In contrast, the floor of the western chamber was noticeable for the dense surface accumulation of shell, much of it burnt, and numerous artefacts. In 1992 three 50 x 50 cm test-pits (HC1, HC2 and HC3) were excavated within the eastern chamber (Figure 9). These were followed in 1994 by a fourth 50 x 50 cm test-pit (HC5), adjoining the western face of Square HC2. A single 1 x 1m test-pit (HC4) was excavated in the western chamber in 1992.

As at Noala Cave, all deposit was passed through nested 3 mm and 6 mm sieves and residues were bagged and returned to the laboratory for sorting, identification and analysis.

Stratigraphy and sedimentary analyses

The eastern chamber excavations revealed a yellow to grey uniform sandy matrix that reached bedrock at between 25–30 cm below surface level (Figure 9). The deposit contained a low density of shell and mammal bone and several pieces of calcrete and volcanic debitage.

In Square HC4 of the western chamber excavations revealed upper layers (Spits 1-5) consisting of yellow to grey sandy strata containing a moderately dense assemblage of marine fauna. This overlay a dark-brown humic lower layer (Spits 6-13) containing very dense midden and a wide range of terrestrial fauna. Bedrock was encountered at varying depths up to 40 cm below surface level. The marked colour contrast in the lower portion of HC4 is due to the higher organic content of the lower layer and the ingress of pale sediments onto the surface of this square.

The pH varied from 8.5 to 9.5 through all strata encountered in Hayne's Cave.

Mineralogical analysis of the sediments (A. Chappell, 1994) indicates a gradual decrease in smectite and a concomitant increase in kaolinite and authigenic aragonite through time, probably reflecting the process of marine transgression and the ultimate proximity of the coastline.

Dating

Three early Holocene dates were returned on shell from Squares HC2 and HC3 (eastern chamber), all of which overlap at two standard deviations (Table 1).

A total of five dates on shell (all aragonite secreting species and XRD tested) and one on charcoal recovered from Square HC4 (western chamber) returned early Holocene dates; in most cases these radiocarbon determinations overlap at two standard deviations (Table 1).

The uncalibrated radiocarbon dates obtained from Square HC4 (n=6) and the other test pits in the adjacent chamber (n=3; see Table 1), corrected for the Oceanic Reservoir Effect, indicate that the Hayne's Cave complex was occupied from approximately 7 800–7 000 BP, commencing around the time Noala Cave was abandoned. While there are some minor inversions apparent in the dates from HC4, a combination of consistent trends in mineralogy and faunal assemblages (see below) through time support our argument for stratigraphic integrity. The detailed analyses of recovered materials presented below refer exclusively to material recovered from Square HC4.

Stone, bone and shell artefacts

A total of 43 stone artefacts were recovered from HC4A and 87 artefacts from HC4B, distributed fairly evenly through the deposit (Data for HC4A in Table 10). A full description of artefacts from this square and the subsequently excavated Square HC4B is provided in Appendix 2. Like those from Noala Cave, the Hayne's Cave artefacts can be separated into two groups based on lithology and likely source. The first group is manufactured from the weakly silicified calcrete that was probably sourced from Barrow Island or a similar bedrock. With the exception of a single multi-platform core, all of the calcrete artefacts are small flakes or flake fragments. The second group of stone artefacts are made on exotic volcanic and metamorphic rocks; these include a flake, flake fragment and a single platform 'horse-hoof' core. Although the sample size is small, it is interesting to note that the volcanic and metamorphic artefacts *only* occur in the lower part of the deposit, while artefacts made of calcrete *only* occur in the upper five spits of the sequence; precisely the time when connections with mainland supply zones may be predicted to have been lost.

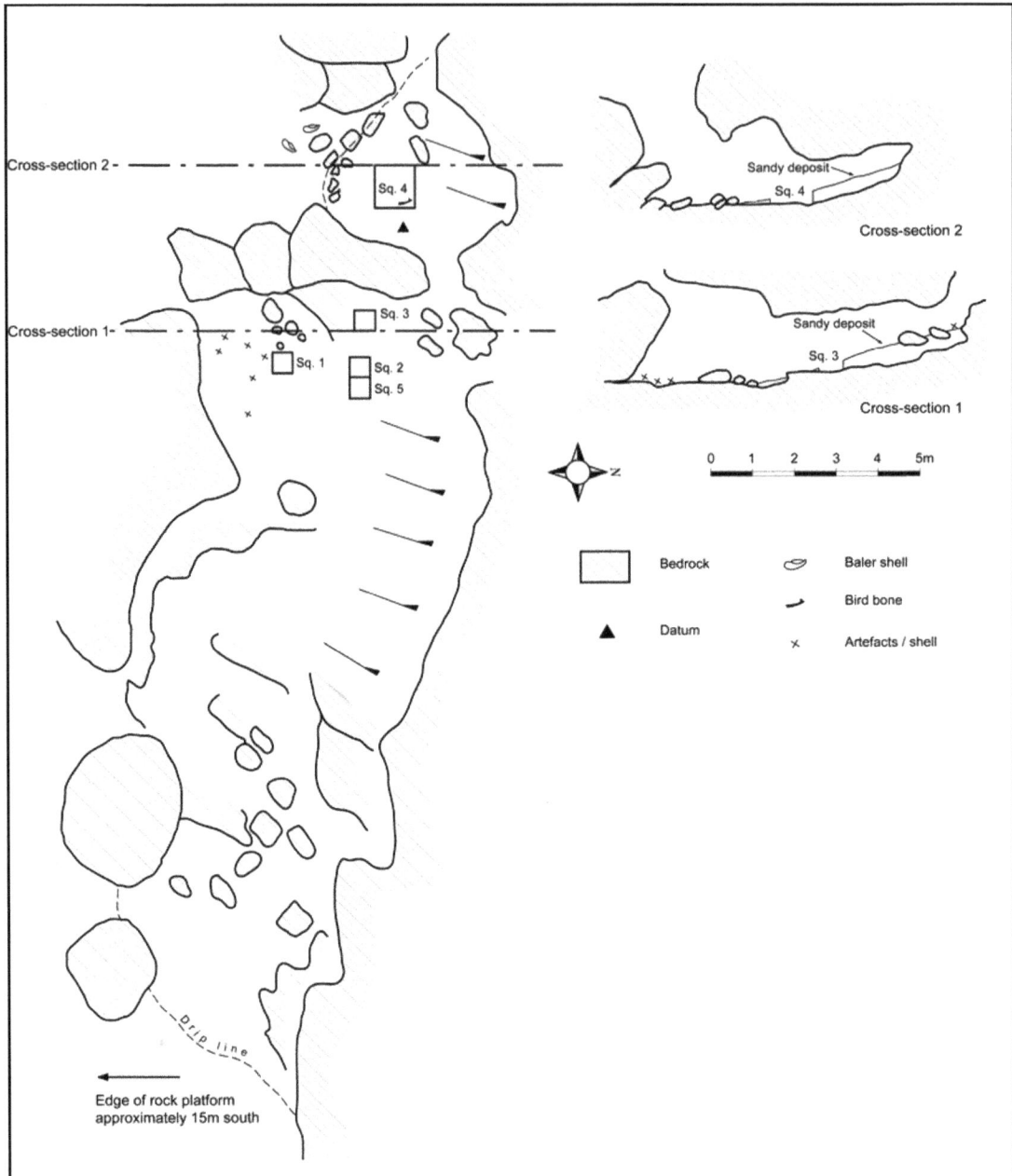

Figure 9   Hayne's Cave plan of test pits and cross section.

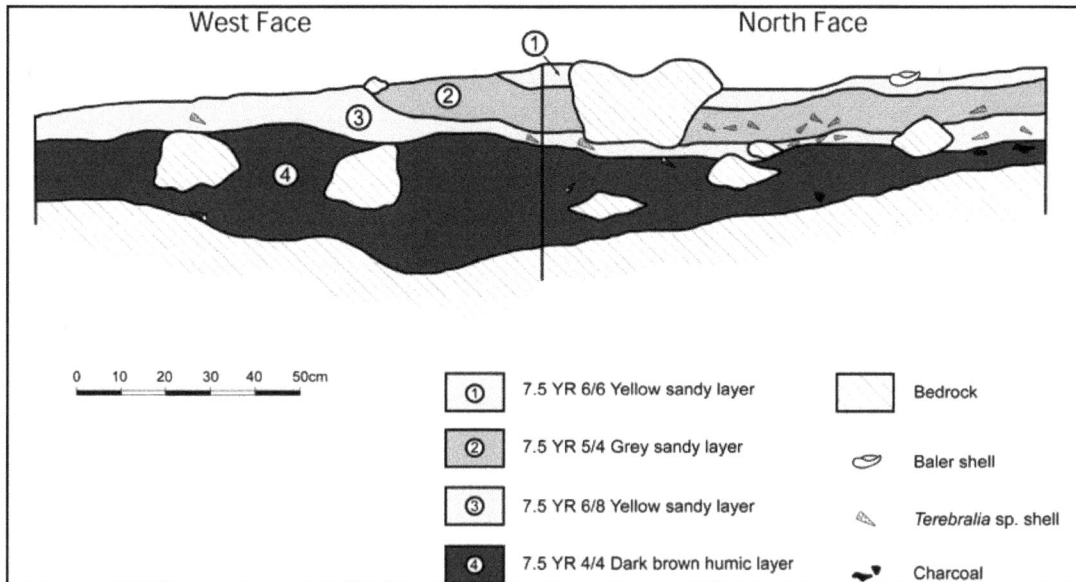

Figure 10   West and north sections of Hayne's Cave, Square HC4.

A single bone artefact in the form of a small bone point with a polished facet and a denticulate edge was recovered from Spit 3 of HC4. Scurla (1996:57-59) describes kangaroo bone points and teeth barbs having been hafted to long wooden spears with resin, for use as composite fishing spears in this region during the ethnographic period.

Four specimens of the predatory gastropod *Natica sagittata* recovered from Spits 3, 6, 7 and 9 have holes in the whorl (Figure 11), suggesting they have been punched through and may have once formed part of a more extensive ornamental shell 'necklace' of the kind described by Morse (1993b) from Mandu Mandu Creek rockshelter on North West Cape and by Balme (2000; Balme and Morse, 2004) from Riwi in the southern Kimberley. It is likely that the hole in at least one of these specimens was created by a carnivorous snail boring into the shell with its rasp-like radula (John Collins, pers. comm.); however, the other three holes are much larger and more irregular, suggesting the carnivorous snail was not responsible for their production.

Shellfish, crustacean, echinoderms and barnacles

The methodology and techniques used by Pulsford (1994) for the analysis of the Hayne's Cave faunal and shellfish remains were broadly consistent with those employed by Manne (1999) for Noala Cave, as detailed above. All samples (or representative sub-samples) of remains were quantified by counts and weights following Bowdler (1983) and Grayson (1984), although the small sample sizes and short time span represented by the deposit restricted the types of numerical analyses that could be carried out. Identifications of molluscs were confirmed by staff at the Western Australian Museum, who also identified sea urchin tests and

spines and small (presumably) non-economic species. Crustacean shell fragments and barnacle fragments were identified by staff of the Museum of Tropical Queensland.

Note that some of the shell (and bone) recovered from Hayne's Cave exhibited surficial crystalline growth, this being either a calcine coating typical of limestone shelters or a soluble calcium carbonate salt precipitant common in arid areas (Lyman, 1994:42). This growth was physically scraped from the more robust specimens (taking care not to damage the underlying shell or bone), but such cleansing was not attempted for the more fragile specimens as the growths, while causing splintering, also provide a level of short-term protection from further splintering. While there are chemical means available for cleaning and stabilising the affected specimens, there is also a real possibility of 'over-cleaning' resulting in their complete destruction (Hope, 1983) and for this reason such techniques were not employed.

MNI values for each economic species were derived primarily from material recovered from the 5 mm sieve fraction owing to the high level of fragmentation of the 3 mm fraction. Table 11 summarises the actual and volumetrically adjusted MNI values for all economic shell species from Square HC4, while Table 12 provides the actual and volumetrically adjusted weights for the same material. Table 13 provides MNI values for non-economic shellfish species.

While there is a wider range of species represented amongst the Hayne's Cave shellfish assemblage than at Noala Cave, again the assemblage is overwhelmingly dominated by *Terebralia* spp. All of the shellfish identified from HC4 with the exception of the mud whelk (*Telescopium telescopium*) are known to occur around the Montebello Islands today (Wells *et al.*, 1994). As with the more restricted Noala Cave assemblage, the Hayne's Cave shellfish remains can be shown to predominantly represent two major near shore habitats: the rocky substrate environment and the mangrove/intertidal zone.

Rocky substrate, shallow water species
High energy rocky shorelines, typically with a double erosion notch, are the most abundant marine environment throughout the Montebello Islands (Wells *et al.*, 1994). A number of economic genera/species from the Hayne's Cave archaeological assemblage can be attributed to this habitat, including *Acanthopleura* spp. (chitons), *Nerita undata* (nerite), *Turbo cinereus*, *Cabestana tabulate* (triton) and *Saccostrea* (oyster; see Tables 11 and 12), of which nerite and triton were not recovered from Noala Cave . *N. undata* is a small gastropod (up to 40 mm) common to rock platforms of the middle and upper intertidal zone,

and fragments of it are present in small quantities in Spits 1–4. *Cabestana tabulate* is represented by a single specimen in Spit 5.

Based on their gross shell morphology, all chitons were initially identified as belonging to the species *Acanthopleura spinosa*. However, subsequent examination of reference material from the sympatric species *A. gemmata* indicates that at least some of the archaeological specimens belong to the latter species. Both species are known to occur in the Montebello Islands, occasionally in mangroves and more typically along intertidal rocky shores, with the latter habitat favoured by *A. spinosa* (Wells *et al.*, 1994).

In general the oyster remains were highly fragmented, a condition which amplified more general species identification problems within this genus and hence such remains were simply attributed to *Saccostrea* sp.

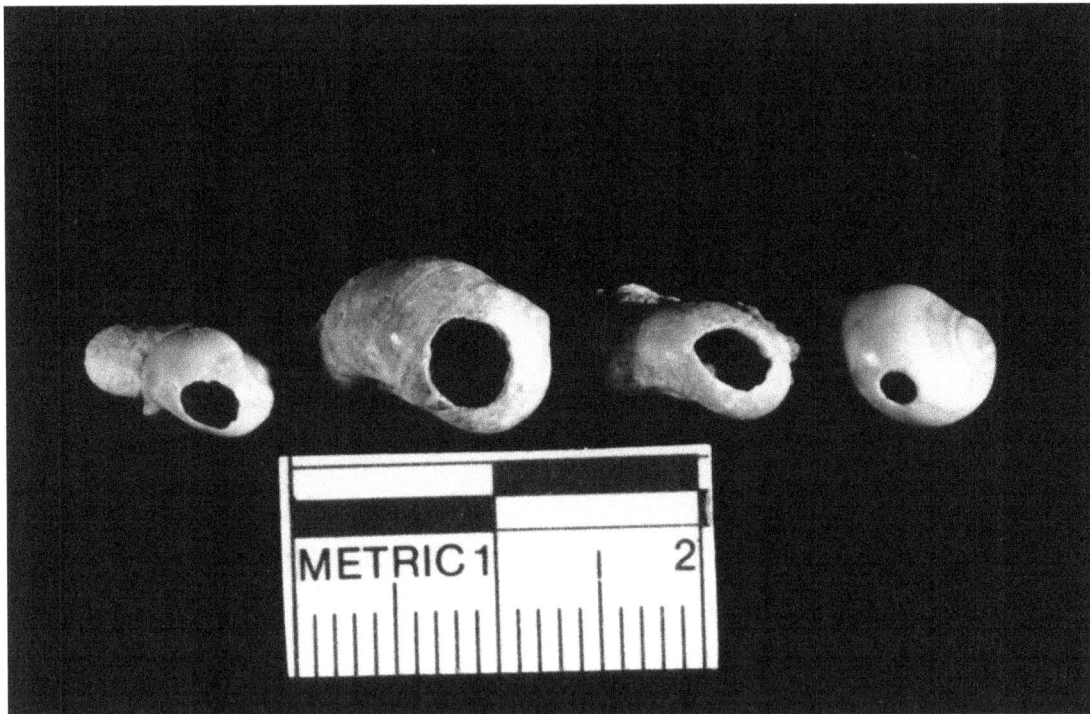

Figure 11   Specimens of Natica sagittata with holes from Square HC4.

Table 10   Stone artefact data, Hayne's Cave.

Haynes Square 4A	Number	%	Silicified limestone	Volcanic
1				
2	3	16	flakes and flake fragments	
3	7	14	multi-platform core, flakes and fragments	
4	6	11	flakes	
5	4	5	flakes and flake fragment	
6	1	5	flake fragment	
7	2	10	flake and flake fragment	flake fragment
8	4	10	flakes	
9	4	7	laterally broken flake	single platform core and flake
10	2	5	flake and flake fragment	
11	4	10	flakes and flake fragments	
12	1	2	flake	
13	2	5	flake and flake fragment	

Mangrove and intertidal mudflat species
Sheltered embayments fringed by mangroves and characterised by fine, soft sandy substrates with low organic content are also comparatively common in the Montebello Islands (Wells *et al.*, 1994). The ubiquitous mangrove genera *Terebralia* spp. (dominated by *T. palustris*, with lesser quantities of *T. sulcata* or *T. semistriata* also present) and *Telescopium telescopium* (mud whelk) are present throughout the deposit.

Other economic shellfish that occur in limited numbers in the Hayne's Cave assemblage include *Rhinoclavis* sp., *Paphia* sp., *Anadara* sp. and *Asaphis violascens*. None of these were recovered from Noala Cave and all four are represented by single specimens only. *Rhinoclavis* is a member of the creepers and is commonly found in sand on tidal flats. The single *Anadara* (mud ark) specimen probably belongs to the species *A. granosa,* an abundant species in mangrove muds of the intertidal zone in northern Australia. *Asaphis violascens* is a bivalve species that is common in sand and gravels of the upper tidal zone, particularly in rocky areas associated with mangroves (Wilson, 2002:70). *Paphia* sp. is a member of the bivalve venus shells, which are commonly found in the sands of the intertidal and subtidal zones of northern Australia.

Non-economic shellfish
Although not extensive in quantity, there is a wide range of non-economic shellfish species also represented in the Hayne's Cave assemblage (Table 13). All of the identified limpet specimens (families Siphonariidae and Acmeidae) are very small and for this reason are considered to belong to this category, rather than being economic.

Crustacea
Small fragments of crustacean carapace were also recovered throughout much of the deposit (Table 14), amongst which three 'taxa' of crab can be recognised: (1) xanthids (Xanthidae), intertidal browsing crabs (typically with a carapace measuring between 1–15 cm across); (2) a species of 'slimy cray' (which are commonly associated with corals) with a maximum length of 10 cm; and (3) significantly, the mud crab *Scylla serrata,* a species normally associated with tropical mangrove habitats and one that thrives in water temperatures between 14–20°C (Hill, 1974).

Echinoderms and barnacles
Small quantities of highly fragmented sea urchin and barnacle remains were also recovered throughout the HC4 deposit (Table 14); owing to their poor state of preservation these have been assigned at the Family level as Echinoidea and Cirrepedia, respectively.

Terrestrial snails
A number of land snails were recovered and identified as belonging to the genera *Rhagada* sp. and *Quistrachia* sp., the highest concentrations of which occurred in Spits 1–5, Square HC4 (Table 14). Like the specimens from Noala Cave, these are not considered to be economic and only have limited use as environmental indicators.

Vertebrate fauna
The abundant vertebrate faunal remains from HC4 were analysed in detail, with considerable effort placed on the identification of both cranial and post-cranial specimens. In general the bone is well-preserved, although, as mentioned above, soluble salts have encrusted some specimens, especially in the lower levels. As in Noala Cave, the assemblage is of mixed origin and includes a significant owl pellet and/or kestrel component in all levels of the deposit.

Fish
While local ethnohistorical sources suggest that fish (and turtle; see below) were an important part of the diet of peoples occupying the North West Cape (Tindale, 1974; Scurla, 1996; Bates, n.d.), they are notoriously under-represented in archaeological sites owing to factors such as post-depositional taphonomy and human discard behaviours (e.g. Casteel, 1976; Colley, 1986). The use of fine meshed sieves during the Montebello excavations mitigated the loss of fish bone data. This approach was rewarded by the recovery of teleost fish bone from all levels of Square HC4 (Pulsford, 1994). Teeth and dentigerous bones were identified by David Bellwood (James Cook University) to Family or Genus level. A number of reef fish were recognised, namely Labridae (wrass and tusk fish; e.g. *Choerodon* spp.); Scaridae (parrotfish; e.g. *Chlorurus* sp. and *Scaras* sp.) and Sparidae (bream). Measurements of vertebrae and teeth were used to estimate the size of individual fish and to calculate more accurate MNI values per spit; the resulting MNI estimates are provided in Table 15.

Parrotfish are reef-dwelling fish of between 30 and 50 cm found in shallow and deeper waters, from where they can be taken using either barbed spears or nets (Allen, 1997; Leach and Davidson, 2000:415). Ethnohistorical sources describe Aboriginal people using the former method to catch fish (Scurla, 1996).

Most of the fish represented in the deposit probably fell in the size range 10–30 cm, although some individual fish exceeded 50 cm in length. The smaller of the fish remains are within the prey

Table 11   *MNI values of economic marine shell recovered from Square HC4, Hayne's Cave.*

	Spit 1	2	3	4	5	6	7	8	9	10	11	12	13	Total
Acanthopleura spinosa	4	15	23	15	14	10	9	4	8	9	7	3		121
Terebralia palustris	37	80	140	63	96	52	82	38	56	74	59	32		809
T. ?semistriata / sulcata	1	3	4	5	2		2	1	3	3	1	2		27
Telescopium telescopium	1			1	1	1	1							5
Nerita ?undata	1	1	3	4	10	1	2	2				1		25
Turbo cinereus	1			3	7	4	9	2	2	4				32
Melo ?amphora	1	2	1		1	1	1		1	1				9
Cabestana ?tabulata						1								
Saccostrea sp.	1	1		1	3		1		1				3	11
Rhinoclavis vertagus										1				1
Anadara sp.							1							1
Asaphis violascens							1							1
Paphia sp.							1							1
Total	47	102	171	92	134	70	110	47	71	92	67	38	3	1044
Volumetrically adjusted total	75	103	173	139	222	159	237	100	102	150	127	63	19	1669

Table 12   *Weights (g) of economic marine shell recovered from Square HC4, Hayne's Cave.*

	Spit 1	2	3	4	5	6	7	8	9	10	11	12	13	Total
Acanthopleura spinosa	24.70	88.10	186.88	116.20	123.32	79.41	79.47	36.49	9.26	60.77	53.08	21.76	1.13	880.57
Terebralia palustris	322.11	836.00	1671.50	993.30	1115.30	653.70	933.84	454.87	648.20	793.67	595.03	372.78	39.30	9429.60
T. ?semistriata / sulcata	0.86	0.37	18.62	10.90	9.40		1.71	2.27	22.95	9.57	5.92	11.93		94.50
Telescopium telescopium	6.01	6.00	4.02	17.91	16.60	2.89	16.85				1.77			72.05
Nerita undata	0.63		1.70	3.40										5.73
Turbo cinereus	0.40	1.00	9.68		27.54	6.84	21.50	3.77	4.59	7.16			6.47	88.95
Saccostrea sp.	3.43	4.00	10.75	10.80	4.40	3.82	9.76	0.00	4.37			1.69	0.14	53.16
Melo ?amphora	39.80		148.59	77.50	61.51	53.79	11.51	5.30	4.99	22.60	2.72			428.31
Cabestana ?tabulata					9.63									9.63
Rhinoclavis vertagus										0.80				0.80
Anadara sp.					1.50									1.50
Asaphis violascens							6.70							6.70
Paphia sp.							2.74							2.74
Total	397.94	935.47	2051.74	1230.01	1369.20	800.45	1084.08	502.70	694.36	894.57	658.52	408.16	47.04	11074.24
Volumetrically adjusted total	635	947	2052	1856	2272	1796	2336	1070	993	1237	998	678.21	325	17195

Table 13   *MNI values of non-economic shellfish, Square HC4, Hayne's Cave.*

Family	Genus / Species						Spit								Total	3mm	6mm
		1	2	3	4	5	6	7	8	9	10	11	12	13			
GASTROPODA																	
Siphonariidae (limpets)		3	19	21	1	2	2	2	10					1	61	42	19
Acmaeidae (limpets)																	
	Patelloida saccharina		5	3	7										15	11	4
	P. mimula		4	4	4										12	12	
Ellobiidae																	
	Auriculastra sp.			1	2										3	2	1
	Melampus sp.									1					1	1	
	Marinula sp.			3											3	3	

30

Taxon	56	88	99	27	18	25	21	6	4	13	8	20	12	16	2	Total	Total
?Laevidentaliidae																	
?Laevidentalium sp.															1	1	
Columbellidae																	
?Dentimitrella sp.	3														3	3	
Turbinidae																	
Astralium sp.		2													2	2	
Turbo cinereus		1													1	1	
Turbo fallaceus		1			7										8	1	7
Turbo sp. (operculum)	3	4	5	2	10	1	2	2						1	33	21	12
Turbo sp. indet.	1				1										1	1	
Cerithiidae																	
Cerithium echinatum		1													1		1
Columbellidae																	
?Dentimitrella sp.	3														3	3	
Trochidae																	
Monodonta labio		1													1		1
?Isandra sp.		2													2	2	
Cypraeidae (cowries)	1	2	1												4	2	2
Naticidae																	
Natica ?sagittata		1	1	1	1		2		1						7	5	2
Gastropod indet.	24	25	11	6	6	1	5	4	1	1	7	5	7	5	119	114	5
BIVALVIA																	
Cardiidae																	
Acrosterigma sp.	3		1												1	1	
Fragum ?erugatum	3	3													3	3	
Fragum sp.															3	3	
Hemidonacidae																	
Hemidonax sp.	2	1													3	3	
Limidae																	
Limatula sp.			1												1	1	
Isognomonidae																	
Isognoman sp.												1			1	1	
Mesodesmatidae																	
Paphies sp.		1													1	1	
Mytilidae																	
Brachidontes sp.		2	2												4	4	
?Septifer sp.														1	1	1	
Tellinidae																	
Tellina sp.		1													1	1	
?Semelidae		1													1	1	
Nueulanidae		1													1	1	
Ungulinidae															1	1	
Numella sp.		1													1	1	
Veneridae																	
Iruss sp.	1	1													2	2	
?Pitar sp.	1	1													2	2	
Tapes sp.												1			1	1	
Hipponicidae	1														1	1	
FORAMINIFERA																	
Marginopora vertebralis	10	18	27	21	6	25	51	4	13	8	20	12	16	2	92	92	
Totals by spit	56	88	99	27	18	25	21	6	4	13	8	20	12	16	2	401	342 / 59
% of total by spit	14	22	25	13	6	6	5	1	1	3	2	5	3	4	1	99.98	85.29 / 14.7

Table 14   Weights (g), NISP and/or MNI values for miscellaneous fauna recovered from HC4, Hayne's Cave.

Spit		1	2	3	4	5	6	7	8	9	10	11	12	13	Total
Miscellaneous fauna weights (g)	Crustacean	0.12		0.23		0.24	0.27	0.72	0.46	0.69	1.18	1.16	0.89	0.16	6.00
	Sea urchin?	1.97	0.96	0.34	1.49	1.94	0.51	0.91	0.29	0.72	1.46	0.26	0.15	0.46	9.61
	Terrestrial snail	1.61	5.63	2.96	4.04	1.81	0.27	0.38	0.75	0.77	0.56	0.34	0.29	0.02	19.79
	Bird eggshell	0.21	2.98	1.76	1.26	0.16		0.06	0.02	0.06	0.01	0.03	0.08	0.02	8.05
	Reptile		0.04	0.27	0.04	0.63	0.31	0.27	0.31	1.52	0.85	0.98	1.11	0.29	6.83
Total weight		3.91	9.61	5.56	6.83	4.78	1.36	2.34	1.83	3.76	4.06	2.77	2.52	0.95	50.28
Volumetrically adjusted total		6.24	9.73	5.56	10.31	7.94	3.05	5.04	3.89	5.38	5.62	4.03	4.18	6.57	77.54
Terrestrial reptile (NISP)	Squamate	9	2	10	3	26	18	9	24	40	45	45	40	13	284
Volumetrically adjusted total		14	2	10	5	43	40	19	51	57	62	66	66	90	525
Terrestrial snail (MNI)	*Rhagada* sp.	1			1	1		1	2	1	4	1			11
	Quistrachia sp.	62	90	9	57	18	3	10	8	17	23	1	13		311
Total		63	90	9	58	19	3	11	10	18	27	2	13		322
Volumetrically adjusted total		101	91	9	88	30	7	24	21	26	37	2	22		458
Miscellaneous fauna presence/absence	Small Bird	*	*	*	*	*	*			*	*	*	*		
	Medium Bird		*	*	*				*	*	*	*	*		
	Small Scincid	*		*						*	*	*	*		
	Medium Scincid											*	*	*	
	Small Agamid			*			*	*	*	*	*	*			
	Small Gekkonid								*		*				
	Small Varanid									*					
	Small Boid								*	*					
	Small Snake			*		*		*			*		*		
	Medium Snake						*					*			

size range of various avian predators—such as the osprey and reef heron (*Egretta sacra*)—and therefore it is possible that some of these derive from bird activity in and around the cave complex. It should be noted, however, that a significant proportion of the fish remains shows signs of burning and hence we favour the view that the majority of the fish remains, particularly the larger ones, are primarily derived from human economic activity. The burnt nature of the fish remains fits will with the ethnographically documented practice of burning fish remains after consumption so as to protect barefooted site occupants from their needle-like bones (cf. Barker, 2004:20).

A wide range of fishing procurement strategies were known from the Pilbara in the post-contact period, including stone fish traps, nets and fishing spears. Given there is no evidence for (surviving) stone alignments or hooks in these deposits the most parsimonious explanation of the fish size ranges described here would be a combination of grass nets and fishing spears (including short reef jabbing spears).

Reptiles
Turtles – Family Cheloniidae
Bone and carapace fragments of large marine turtles were recovered from all spits bar the uppermost two, with peak concentrations in Spits 3, 8 and 9. Only a small proportion of the turtle remains could be identified to anatomical units, but these include fragments of carapace, caudal vertebrae and bones from the flippers. The absence of turtle eggshell in the deposit is probably a result of post-depositional disintegration of this parchment-like material.

These fragmentary remains could not be assigned to species, but they are most likely derived from the green turtle (*Chelonia mydas*) which is the commonest species nesting and feeding in the Barrow-Montebello Island group today (Limpus and Miller, 1993). Adult green turtles generally weigh 90–180 kg and yield at least 60 kg of flesh, making them a highly profitable prey item. Green turtles were observed by MARP members in the water and on beaches during the 1992 field season. Other possible candidates include the flatback turtle (*Natator natator*) and the hawksbill turtle (*Eretmochelys imbricata*), but unlikely given that the former is scarce in the region today and the flesh of the latter is known to be toxic. Turtles occasionally lose their way while searching for a nesting site and can die of thirst or exposure (e.g. observed on Barrow Island). Their partially-articulated remains are sometimes observed on the surface of rockshelters deposits (Ken Aplin, pers. obs.). However, the fact that turtle bone and carapace is present at low but fairly consistent levels through the Hayne's Cave midden points to an economic origin for the remains rather than occasional chance deaths. Scurla's (1996) ethnohistorical account describes people of the Cape Range area hunting turtles from wooden watercraft using spears.

Estuarine Crocodile (*Crocodilus porosus*)
Two crocodile teeth were recovered from Spit 6 and a conjoined tooth fragment from Spit 7 of Square HC4 (Figure 12). The fact that no other identifiable pieces of crocodile bone were recovered raises the possibility that the teeth may have been decorative items rather than the remains of economic activity. Estuarine crocodiles have a strong preference for mangrove communities and they may have been more continuously distributed along the northwest coastline during the early Holocene when this habitat was more extensive.

Terrestrial reptiles (lizards and snakes)
Four families of lizards and two of snakes were identified amongst the faunal remains from Hayne's Cave. Table 14 documents the presence/absence of these remains throughout the HC4 deposit. Unfortunately, the snakes have not yet been examined in detail, so we are unable to compare the species diversity with that of Noala Cave.

As with Noala Cave, the majority of bones from the larger lizards and snakes are fragmented and some are burnt; hence these remains are considered to be economic in origin. In contrast, bones of the smaller sized reptiles are generally unfragmented and only rarely burnt; these are interpreted as most likely deriving from kestrel or owl activity in the cave complex. In general, the remains of small reptiles were far less numerous in Hayne's Cave than in the Noala Cave deposit.

Birds
Bird bone and fragments of thin, unidentified eggshell were recovered in small quantities throughout the deposit (see Table 14). Most of this material is unburnt and undamaged, and is likely to be non-economic; some probably derives from owl or kestrel activity, while other, larger skeletal elements may represent natural deaths in the cave. An almost intact synsacrum of a bird was found on the surface of the deposit.

Mammals
The mammalian fauna from Hayne's Cave (Table 16) are very similar in taxonomic composition to those recovered from Noala Cave, with the exception of three small murids and a variety of small dasyurids which were recorded exclusively in the Noala Cave deposit. This might simply reflect the overall greater quantity of owl pellet remains in the latter site. However, it may be significant that two of the murids which are absent

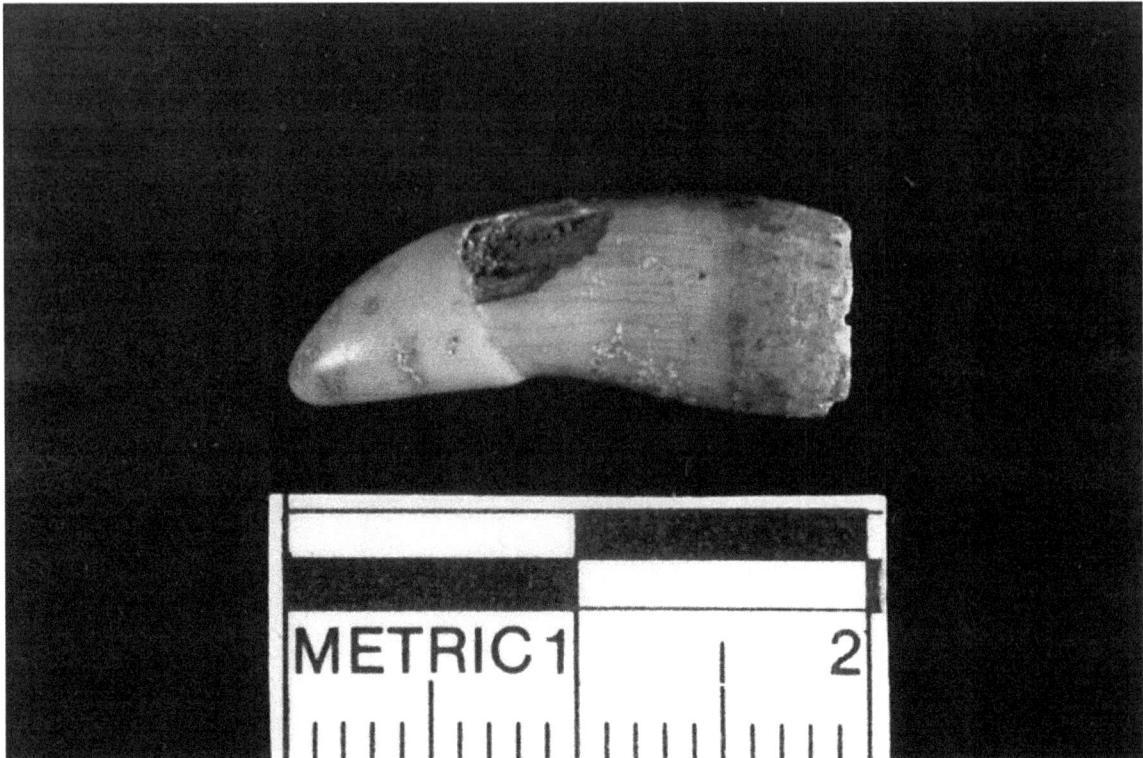

Figure 12   Photograph of salt water crocodile molar.

from Haynes Cave—*Pseudomys fieldi* and *Notomys longicaudatus*—are also confined to the lower spits of Noala Cave. Both are specifically associated with non-sandy substrates. MNI values for the economic species are provided in Table 16. Data on the weights, NISP and MNI values, and the presence/absence of miscellaneous fauna are provided in Table 14.

Marsupial carnivores – Family Dasyuridae
The same two large bodied marsupiocarnivores that were found in the Noala Cave deposit are also present in Square HC4 of Hayne's Cave: the western quoll (*Dasyuuis geoffroii*) and the northern quoll (*D. hallucatus*). The former occurs in Spits 3, 5 and 8–11, while the latter is represented in Spits 6–7. In addition, a further five specimens attributed to this Genus (in Spits 5 and 8–11) could not be taxonomically identified further.

Bilbies- Family Thylacomyidae
A single specimen of *Macrotis* sp. (possibly *M. lagotis*, the greater bilby) was recovered from Spit 11 of HC4.

Bandicoots – Family Peramelidae
As in Noala Cave, three different bandicoots are represented in Hayne's Cave: *Perameles bougainville* (the western barred bandicoot), *Isoodon auratus barrowensis* (the golden bandicoot) and a larger sized unidentified species of *Isoodon*. Interestingly, remains of *P. bougainville* are restricted to the basal two spits in Hayne's Cave; it was also absent from the upper spits of Noala Cave. The larger species of *Isoodon* is sparsely represented in the site (occurring in the lowest two spits), while the smaller Barrow Island form (*I. auratus barrowensis*) is moderately abundant throughout most of the upper deposit.

Table 15   MNI values for marine fish bone recovered from Square HC4, Haynes Cave.

Spit	1	2	3	4	5	6	7	8	9	10	11	12	13	Total
Scaras sp.	2	2	1	3	1	1	1	1	1	1		4		18
Chlorurus sp.									1				1	2
Choreodon sp.			1	2		1	1		2	1				8
Sparidae					1				1		1	1		4
Labridae									1		1			2
Total	2	2	2	5	2	2	2	3	4	2	2	5	1	34
Volumetrically adjusted total	3	2	2	8	3	4	4	6	6	3	3	9	7	60

Possums – Family Phalangeridae
The northern brush-tail possum (*Trichosurus vulpecula arnhemensis*) is found in small quantities throughout much of the Hayne's Cave deposit.

Rat Kangaroos – Family Potoroidae
Both the larger, southern form and the smaller, Barrow Island form of *Bettongia lesueur* are represented in the Hayne's Cave deposit. In contrast to the situation in Noala Cave, the small Barrow Island form is the less abundant of the two; it occurs only sporadically in Spits 2, 3, 5 and 7, although it appears to become relatively more important in the uppermost spits of the site. In contrast the larger form of *B. lesueur* occurs continuously between Spits 3 and 12.

Kangaroos and wallabies – Family Macropodidae
A total of five macropodids are represented in the Hayne's Cave assemblage, one more than in Noala Cave. These include the medium sized macropods *Lagorchestes conspicillatus* (found throughout the deposit except Spit 1), *L. hirsutus* (occurring infrequently) and *Petrogale* sp. (only in Spits 4, 8 and 11), as well as the remains of a wallaroo (*Macropus robustus*) in Spits 5 and 10. The additional species is the northern nail-tail wallaby, *Onychogalea unguifera*, a species of subtropical northern Australia that occurs today northwest of the Montebellos in the vicinity of Broome (Ingleby and Gordon, 1998). Across its extensive range, this species occupies a variety of habitats including woodland with grassy understorey, coastal plains with tussock grasses and thickets of *Melaleuca*, and black soil plains of tussock grasslands with scattered trees and shrubs. The northern nail-tail wallaby is known as a selective feeder that eats mainly dicotyledonous herb foliage, succulents, fruits and green grass shoots (Ingleby, Westoby and Latz, 1989; Ingleby and Gordon, 1998). Previously, this species had not been recorded among subfossil remains from the Cape Range Peninsula or the Pilbara (Baynes and Jones, 1993).

The northern nail-tail wallaby is represented in both the lower and upper deposits of Hayne's Cave (see Table 16), although its remains are less abundant than those of either *Lagorchestes conspicillatus* or *Bettongia lesueur*. The question as to why this species should be absent from the adjacent Noala Cave deposit is intriguing. One possible explanation is that northern nail-tail wallaby populations underwent a southward expansion along the exposed continental shelf during the initial period of transgression, during or after the final phase of occupation of Noala Cave, but prior to first usage of Hayne's Cave. Given the specific requirement of this species for herbaceous and succulent vegetation, its occurrence in the area suggests elevated moisture availability within the coastal dune field during the period of occupation of Hayne's Cave compared with more arid conditions during the terminal Pleistocene. Such a change is perhaps not unexpected during the early phase of transgression when rising postglacial temperatures could be expected to produce a broad expanse of warm, shallow water over the flooded portion of the continental shelf (van der Kaars and De Deckker, 2002).

Bats
Small microchiropteran bats were recovered from several spits in Hayne's Cave, but there was no evidence of either Ghost Bat or Flying Fox remains. Microchiropterans are represented in many owl pellet deposits, but usually in low numbers (Ken Aplin, pers. obs.).

Taphonomic considerations of the Hayne's Cave fauna

Small mammal, bird and reptile remains were recovered throughout the site. As in Noala Cave, these remains are generally not fragmented and only occasionally burnt, and they are interpreted as most likely representing the disaggregated remains of owl and/or kestrel pellets.

The larger reptile and mammal remains are considerably more fragmented and more often burnt, and are identified as economic food remains.

Changing patterns of habitat and resource use within the Hayne's Cave catchment

Each of the MNI and raw weight data for economic species show a general decrease in terrestrial fauna in the Upper Unit of the deposit, particularly in Spits 1–4. This contrasts with the persistence or increase during this last phase of occupation of marine fauna including turtle, fish, shellfish, crustacea and sea urchin, a trend noted previously by Veth (1995). The obvious explanation for this pattern is that terrestrial habitats were being progressively lost as the coastal plain was drowned, with a likely rise in the productivity of marine habitats.

In this section we will examine in more detail some of the evidence for the emergence and exploitation of key environmental habitats during the early Holocene in the area surrounding Hayne's Cave. This evidence has important implications for the hotly debated issues of the relative productivity of transgressive shorelines and the presence or absence of critical habitat types, such as mangrove communities during this period (e.g. Beaton, 1995).

Marine habitats
The four marine environments exploited by the occupants of Hayne's Cave were rocky shores, intertidal mudflats, mangroves and coral reefs. By

Table 16   MNI values of economic mammal bone recovered from Square HC4, Haynes Cave.

Spit	1	2	3	4	5	6	7	8	9	10	11	12	13	Total
Bettongia barrowensis		1	1		1			1						4
Bettongia lesueur			2	1	2	2	1		2	2	2	1	1	16
Dasyurus geoffreii			1		2			1	3	1	2			10
Dasyurus hallucatus						1	1							2
Dasyurus sp.					1			1	1	1	1			5
Isoodon auratus	1	1	2	2	2	2	3	3	4	3	4	3	2	32
Lagorchestes conspicillatus		1	1	1	1	1	1	1	1	1	2	2	1	14
Lagorchestes hirsutus					1				1		1	1		4
Largorchestes sp.	1	1	1		1	1		2	1	3	1	1	1	14
Macropus robustus					1					1				2
Macrotis sp.											1			1
Onychogalea unguifera			2	1	2	2	1	2	3	1	2	2		18
Perameles sp.												1	1	2
Petrogale sp.				1				1			1			3
Trichosurus sp.	1		1	1	1	1	1		1	2	1	2		12
Bettong			1	1	2	1	1	1	1	1	1	3	1	14
Peramelid			2		2	2	1	1	2	3	2	2	1	18
Total	3	4	14	8	19	13	11	15	20	19	20	18	7	171
Volumetrically adjusted total	5	4	14	12	32	27	24	32	30	27	28	31	48	314

the early Holocene, the rate of sea level rise was probably less than 1 m/100 yrs (cf. Chappell and Thom, 1986) with many species either able tocolonise, disperse or regenerate successfully in the face of such slow rates of transgression.

Rocky shores
Exploitation of rocky shoreline habitat is indicated by specimens of the sessile molluscs oyster (*Saccostrea* sp.), chitons (*Acanthopleura* spp.) and possibly limpets (Pattelidae). The life cycle from larval stage to attachment and breeding age for these molluscs is approximately 2 to 5 years (John Collins pers. comm.) so at most sea level rise might have a pruning effect, producing populations with fairly even age profiles through time. Mobile taxa, such as crustaceans, active gastropods and fish would essentially be unaffected.

Intertidal mudflats and unconsolidated sediments
Intertidal mudflats were clearly present in the Montebello Islands during the early Holocene, having been necessary to support viable (although perhaps spatially restricted) populations of *Geloina coaxans*, venus shell (*Paphia* sp.), ark shell (*Anadara* sp.) and *Asaphis violascens*, the remains of which are common in the deposit. This environment is probably the most dynamic during times of rising sea levels (Chappell and Thom, 1977; Davis and Clifton, 1987; Leatherman, 1990) and may have been the least stable and productive, from an economic perspective.

Mangroves
Protected embayments are likely to have developed on the landward (eastern) side of the Barrow-Montebello Peninsula in the early Holocene, providing suitable habitats for the establishment of mangrove communities. The presence of such communities is attested to by the abundance of gastropods such as *Terebralia* spp. and *Telescopium telescopium* in Hayne's Cave. Many of these shells show evidence for both processing and burning, leaving no doubt as to the importance of this zone for the occupants of the site. The presence of mud crabs (*Scylla serrata*) and estuarine crocodile (*Crocodylus porosus*) also provide evidence not only for the existence, but also for the exploitation, of the mangrove habitat at this time. The Montebello Island evidence is consistent with a regional picture of the existence of extensive mangrove stands during the late Pleistocene to early Holocene, as indicated by pollen records from deep sea cores taken to the north of the Sahul Shelf (van der Kaars, 1991; van der Kaars *et al.*, 2000).

Pulsford (1994:75–9) has provided a detailed review of current models for mangrove development in the early Holocene, suggesting three phases in their development around the Montebello Islands. The first phase, dating to the terminal Pleistocene, saw the limestone escarpment situated well inland and protected by a wide belt of sand plains. The coast at this time probably supported a broad, fringing mangrove community that marched steadily landward as the sea level crept even higher. The second period, dating to approximately 6 000 BP, began immediately after the sea attained its present position up against the Montebello 'plateau'. At this time, many large mangrove communities would have collapsed owing to an absence of a substrate suitable for mangrove colonisation. The final phase, commencing after stabilisation of sea level, saw a progressive build-up of autochthonous sediment in more sheltered locations, providing new substrate for a limited increase in the areal extent of mangrove communities around the archipelago once more. Under this model, mangrove resources would have been at their maximum during the period of marine transgression, provided that this did not outstrip the rate of lateral movement of the mangrove community through colonisation.

Coral reefs
The remains of large parrot fish (*Scaras* sp.) and wrasse (*Chlorurus* sp.) in the site deposits

suggests that a fully developed fringing reef structure was present around the Montebello coast (David Bellwood, pers. comm.). This inference is supported by observations on the history of coral reef structures at the Houtman Abrolhos Reef to the south. Here, coral growth had commenced as early as 9 500 BP with the colonisation of Pleistocene substrates (Collins, Zhu, Wyrwoll, Hatcher, Playford, Chen, Eisenhauer and Wasserburg, 1993). By approximately 9 000 BP, water depth around these near-shore continental islands was approximately 20 m, allowing corals to rapidly colonise the adjacent rocky slopes (Hopley, 1984).

Terrestrial landscapes
The majority of the terrestrial habitats that once surrounded Hayne's Cave are now drowned. However, the more extensive and topographically complex landscape of nearby Barrow Island provides some clues as to the character of the Barrow-Montebello Peninsula during periods of lowered sea level. Two broad habitats are discussed here, with reference to the mammal fauna recovered from Hayne's Cave.

Low dissected limestone plateau
On Barrow Island the low dissected limestone plateau is dominated by sparse to dense hummock grassland of *Triodia* spp., low shrubs including *Acacia bivenosa* and *Cassia* spp. and low clumps of the native fig, *Ficus platypoda*. These habitats support moderate to high densities of the spectacled hare wallaby (*Lagorchestes conspicillatus*), the Barrow Island bettong (*Bettongia* sp.) and the wallaroo (*Macropus robustus*) (Short and Turner, 1991). Other rock dwelling species that might be expected to occur around the margins of this habitat include rock wallabies (*Petrogale* sp.), the northern quoll (*Dasyurus hallucatus*; note that this species is absent from Barrow Island today) and the common rock rat (*Zyzomys argurus*) (Watts and Aslin, 1981; Bradley, Kemper, Kitchener, Humphreys, How and Schmitt, 1988; Braithwaite and Begg, 1998). In addition, brush-tailed possums (*Trichosurus vulpecula*) shelter among rocks and crevices, feeding primarily on the leaves and fruit of *Ficus platypoda*. The golden bandicoot (*Isoodon auratus barrowensis*) is common across all habitats on Barrow Island, taking shelter by day in large *Triodia* hummocks and among rocks.

Sand plains
Sand plains and associated dune field habitats would have occupied large areas of the exposed continental shelf, much of it probably mantled by *Triodia* hummock grassland. As sea level rose ever higher, these habitats would have decreased progressively in both area and continuity. During the time period represented by the lower half or more of the Hayne's Cave deposit, terrestrial mammals derived from these habitats were of great significance. Economic species recovered from Hayne's Cave that we identify as most likely coming from the sand plain habitat include the long-nosed bandicoot (*Perameles bougainville*), the western quoll (*Dasyurus geoffroii*) and several small to medium-sized macropodids, namely the burrowing bettong (*Bettongia lesueur*) and the rufous hare wallaby (*Lagorchestes hirsutus*) (Baynes and Jones, 1993; Friend, 1990; various authors in Strahan, 1988). The northern nail-tail wallaby (*Onychogalea unguifera*) almost certainly occupied the coastal plain habitat, probably concentrating around low-lying areas such as clay pans where there was sufficient moisture retention to support a herbaceous ground cover. The absence of this species from the slightly earlier deposit in Noala Cave suggest that such conditions were probably absent or more restricted prior to *circa* 8 000BP, and that conditions for human exploitation of the coastal plain habitat were optimal during the brief period of approximately 500–800 years during the initial phase of marine transgression. However, as the coastline advanced, large areas of the more productive central plain habitat were initially replaced by resource rich mangrove communities and then lost altogether, resulting in a comparatively depauperate coastline and hinterland by the mid-Holocene.

In comparison with other regional faunal assemblages, the suite of mammals from the Hayne's Cave deposit is remarkably diverse. This is particularly striking when comparisons are made to the extant Barrow Island mammal fauna. However, it is also true when comparisons are made with the 'original' mammal fauna of the Cape Range Peninsula to the south. In the latter case, the major differences relate to the diversity of medium-sized mammals, the most notable additions being one of the two *Bettongia* species (one taxon was present on the Cape Range Peninsula, but it is unclear as to exactly which species), and three macropodids: *Onychogalea unguifera*, *Lagorchestes hirsutus* and *L. conspicillatus*. Most of these additional species are likely to have been elements of a now-submerged coastal plain fauna. Significantly, only one of these species has survived to the present on Barrow Island. This is the spectacled hare wallaby (*L. conspicillatus*), a taxon that is also able to survive in areas of rocky limestone plateau owing to its extreme physiological adaptations for water retention.

Several taxa identified in the Cape Range deposits by Baynes and Jones (1993) were not recorded amongst the Montebello Island deposits. These include two moderately large murids, a Stick Nest Rat (*Leporillus apicalis*) and the Golden-backed Tree Rat (*Mesembriomys macrurus*). The latter taxon is closely associated with tree cover across its extensive but patchy range, and its apparent absence from the habitat is suggestive of an essentially treeless Barrow-Montebello Peninsula. *Leporillus apicalis* might have occurred in low numbers around the rocky limestone plateaux and simply been missed during excavation. Alternatively, this essentially southern taxon may have been at the northernmost limit of its range at North West Cape.

Two additional large mammals have been recorded from basal, late Pleistocene levels in archaeological sites on the Cape Range Peninsula, these being the agile wallaby, (*Macropus agilis)* and the Tasmanian Tiger or Thylacine (*Thylacinus cynocephalus*) (Baynes and Jones, 1993; Morse, 1993b; Przywolnik 2002b). These taxa may have already disappeared from the area prior to the onset of accumulation of the Noala Cave deposit, or they might have been too scarce to be recorded in the volumetrically small recoveries from this site. The Echidna (*Tachyglossus aculeatus*) is also recorded in both the modern and cave faunas of the Cape Range Peninsula. Its absence in both Montebello Island cave deposit faunas most likely indicates its relative scarcity on the Montebello-Barrow Peninsula throughout the late Pleistocene to early Holocene period. Baynes and Jones (1993) have suggested that the remains of *M. agilis* from the basal layer at Mandu Mandu Creek rockshelter might actually predate the first archaeological material by some unknown period. The absence of this taxon from either of Noala or Haynes Caves lends support to this view. The southernmost record of *M. agilis* today is from the vicinity of Mandora, near the southern margin of the Great Sandy Desert (Peter Kendrick, pers. comm.). This population appears to be relictual and disjunct from the main distribution in the north and it may be a remnant of a formerly more extensive distribution of agile wallabies along the coastal strip

Hayne's Cave: An early Holocene coastal economy

We envisage that around approximately 8 000 BP the prehistoric occupants of the Montebello area shifted their residential preference from Noala Cave, perhaps made less attractive by the increasing proximity of the sea, to the better sheltered cave/chambers of the Hayne's Cave complex. At this stage the various islands were still largely connected to each other and perhaps also to Barrow Island, and separated by only a 5 km wide passage to the mainland. Cessation of occupation in Hayne's Cave by approximately 7 000 BP may be reasonably assumed to signify abandonment of the islands by people as they became the far flung outliers of their contemporary configuration, particularly given the absence of any open surface sites. No pre-contact cultural material has been located during either excavation of systematic surface survey that might be reliably dated to after 7 000 BP. The extreme

isolation of the small islands and their lack of 'connectivity' to Barrow Island and the mainland would in our view have rendered them eventually a biogeographic dead-end.

The Hayne's Cave assemblage provides a unique insight into the early Holocene subsistence strategies of people utilising the now-drowned northwest shelf. As detailed above, the abundant faunal remains document the exploitation of a wide variety of both terrestrial and marine habitats. On the land, mammals and other vertebrates were obtained from sand plain, dune field and limestone plateau habitats, while the diverse littoral environment of fringing reef, mangroves and rocky shores provided an extensive range of marine resources ranging from turtles to medium- and small sized fish, crab, lobster and a host of shellfish species. The lower spits of Hayne's Cave in particular contain a wide range of terrestrial and marine fauna, indicating both proximity to the sea and access to extensive areas of relatively productive terrestrial habitats. In contrast, the paucity of terrestrial fauna in the upper layer is consistent with a gradual loss of the coastal plains habitats through encroachment of the sea across the broad continental shelf.

The presence of metamorphic and volcanic lithologies among the relatively sparse stone artefacts recovered from both Noala and Hayne's Caves indicates some degree of contact with areas over 100 km distant on the contemporary mainland. This implies either that exchange down-the-line was operating in the terminal Pleistocene and early Holocene, or that single bands or groups had very large territories incorporating significant portions of the 'interior'.

Perusal of 'tribal'—or more correctly linguistic—boundaries for the northwest of Australia from Tindale (1974) and other sources shows that the largest interior territories are indeed associated with groups abutting deserts, such as the Great Sandy Desert, which lack co-ordinated drainage. In contrast, those deserts that are part of major catchments, such as on the Abydos Plain of the Pilbara, have the smallest interior territories. As with the earlier prehistoric period covered by Noala Cave, a high degree of residential mobility is implied.

Owing to a range of sampling and taphonomic issues discussed above, it has not been possible to perform a meaningful quantitative analysis of the contribution to the diet of terrestrial versus marine resources. Nevertheless, we believe that the observed differences between the faunal assemblages from the upper versus lower spits of Hayne's Cave are consistent with the known landscape history and are indicative of a progressive shift in emphasis from terrestrial to marine resources in the vicinity of the site. Should human skeletal remains dating to this time period be recovered from the region in the future, bone isotope studies might shed some light on this particular issue.

At this stage, it may be profitable to consider whether or not the evidence from Hayne's Cave is indicative of a true, coastally-tethered society, or whether it reflects an arid coastal plains adaptation where marine resources are used in a complementary fashion. Such a distinction is of course very difficult to draw on the basis of a single site. Nevertheless, what does seem clear from the Montebello evidence is that a wide range of terrestrial fauna was exploited for as long as it was available, and further, that this fauna was drawn from all of the regionally represented habitats including the formerly extensive coastal sand plain and associated dune field. The presence of 'exotic' lithics also suggests that access was maintained to more inland resources, either through direct visitation or through contact with inland-dwelling populations. Either way, the evidence points to a degree of mobility that is difficult to reconcile with the notion of a coastally-tethered society, and to an economic system in which marine resources played a complimentary rather than a dominant role. On the other hand, it is clear from the combined evidence of Noala and Hayne's Caves that marine resources were immediately and extensively exploited whenever the sea came to within an optimal distance (i.e. 5–10 km) of the sites. Similar conclusions have been reached by other workers based on sparse late Pleistocene to early Holocene assemblages from other sites in north-western Australia (see below); hence, it seems reasonable to conclude that regular and significant exploitation of marine resources was occurring on a regional scale throughout this period. That much of this activity remains archaeologically invisible is probably explained by the excessive distance, through much of the late Pleistocene, between the coastline and the few excavated sites.

Comparison of the Montebello sequences to other coastal sites in the region

While rockshelters are not especially common along the Pilbara coastline, they are present on both the Burrup Peninsula and Cape Range Peninsula where they have been the focus of sustained archaeological investigations. One of the earliest such investigations of a suite of small coastal caves was that by Morse (1988, 1993a, 1993b, 1996, 1999) at North West Cape, where the offshore profile is so steep that Pleistocene shorelines have never been located further than 10–12 km from the sites. The sites of Mandu Mandu Creek, Pilgonaman Creek and Yardie Well collectively provide a record of 30 000 years of occupation of these arid coastal plains with

varying degrees of reliance on marine fauna. Morse (1999:77) concluded:

> Occupation of the Cape Range Peninsula appears to have always focused on the coast and its resources. During the arid conditions of the last glacial period people followed and adapted to the changing Pleistocene coastline making rare visits to the coastal hinterland, and re-emerging in the archaeological record with a comprehensive and sophisticated knowledge of coastal resources as sea level stabilised during Holocene times.

She also drew attention to the fact that evidence for marine resources being an integral part of economies in the terminal Pleistocene now comes from northern Melanesia, the west Kimberley and Shark Bay.

Recent excavations by Przywolnik (2002b, 2005) at the C99 and Jansz rockshelters further north on North West Cape, have also provided Pleistocene-aged sequences showing a human reliance on both marine and terrestrial suites. She noted a hiatus in deposition between approximately 21 000 BP and 10 700 BP, stating that during the LGM '…sediment and artefact accumulation rates show that deposition either stopped completely, or slowed dramatically…This supports the conclusion that human use of the sites at this time all but ceased' (Przywolnik, 2002b:300). The reason for this change in occupation was attributed to a likely combination of movement of groups along a retreating coastline and their relocation to other areas to escape increasing aridity. A third shelter, Wobiri, was also excavated by Przywolnik (2002b, 2005). An (uncalibrated) date of 5 160 BP represents the commencement of human use of this site, although the most intensive use of the site occurs after *circa* 2000 BP. The Wobiri deposits contain a range of stone artefacts, bone and shellfish remains, and while terrestrial species are present, the assemblage is overwhelmingly dominated by marine species including turtle, fish and shellfish.

The Cape Range Peninsula, on which the key rockshelter sites of Morse and Przywolnik are located (Figure 1), would have been directly connected via Barrow Island to the Montebello Islands for much of the time of their occupation. It is not surprising, therefore, that both the terrestrial and marine fauna from the Cape Range sites are similar to those recovered from Noala and Hayne's Cave. Minor differences occur in the mammalian fauna including the presence of agile wallaby (*Macropus agilis*) in the basal Pleistocene Unit of Mandu Mandu Creek and thylacine (*Thylacinus cynocephalus*) in Pilgonaman Creek during the early Holocene. As indicated earlier, the agile wallaby record might conceivably predate the archaeological sequence at Mandu Mandu Creek. Like the northern nail-tail wallaby from Hayne's Cave, *Macropus agilis* is a northern subtropical species that today extends south along the coastal margin of the Great Sandy Desert as far as Mandora (Peter Kendrick, pers. comm.); it does not occur today in the Pilbara uplands or on the adjacent coastal plain. It too may have extended its range south during a prior period of lowered sea level when conditions on the coastal plain were sufficiently mild to support adequate growth of soft tussock grasses.

Excavations in several small rockshelters on the Burrup Peninsula (Figure 1) are yet to yield evidence for late Pleistocene occupation, although they have returned dates as early as 6 750 BP, with evidence for exploitation of mixed marine and arid terrestrial fauna at this time (Vinnicombe 1987). Terrestrial fauna from these sites include the euro (*Macropus robustus*), a rock wallaby (*Petrogale* sp.) and the black flying fox (*Pteropus alecto*). Marine fauna include reef dwelling species of fish (e.g. Sparidae and Labridae), as well as crabs and gastropods derived from mangrove communities. Excavations on various islands of the Dampier Archipelago and surrounds by each of Lorblanchet (1977, 1992), WA Museum staff (Vinnicombe, 1987), Bradshaw (1995), Przywolnik (2002b) and Clune (2002) have identified middens deposited from 8 500 BP onwards. These clearly illustrate the exploitation of marine resources as soon as the sea initially approached the outer islands, the inner islands and then the mainland. Pleistocene ages for some of the copious and heavily altered engravings of the Burrup Peninsula have long been suggested but not yet confirmed (Veth, Bradshaw, Gara, Hall, Haydock and Kendrick, 1993). The lack of Pleistocene dates for this coastal area is almost certainly a product of the lack of major shelters and the intractability of dating the engravings. Clearly groups of people were present on the arid coastal plain throughout the Holocene. Moreover, these groups evidently were highly mobile, exploiting both marine and terrestrial resources.

A number of open midden sites have been studied from various sections of the Pilbara coast, including the Warroora coastline, Cape Range Peninsula, Burrup Peninsula, the Pilbara coastal plain and Eighty Mile Beach (Figure 1). The sites vary from small monospecific scatters and mounds through to large stratified linear and mounded middens, the latter sometimes containing a wide range of shellfish, fish, turtle and occasionally dugong and terrestrial mammals. Often the large middens are found near to the more permanent water sources (Veth and O'Brien, 1986; O'Connor and Veth, 1993). The earliest dated open midden is on the Warroora coastline and dates to *circa* 7 400 BP (Kendrick and Morse, 1982), near the end of the period of occupation of Hayne's Cave. Faunal remains include a wide range of shellfish derived from intertidal flat, mangrove, reef and rocky shore habitats, as well as unspecified marine fish, crab, sea urchin and turtle. An unspecified macropodid was also represented. Another midden, at Mulanda Bluff dates to *circa* 7 200 BP and contains abundant remains of a mangrove-dwelling mud whelk (*Terebralia* sp.). Middens on the Cape Range date to as early as *circa* 5 700 BP and contain baler shell (*Melo amphora*), chiton (*Acanthopleura gemmata*), giant clam (*Tridacna maxima*), turban (*Turbo* sp.) and *Terebralia* sp. (Morse, 1996).

Veth (1995) provides data for other shelters and midden sites which have been excavated on the arid Pilbara coastline, most of which are associated with post-transgressive marine catchments. The earliest of the published sites, Site P2772 on the Burrup Peninsula, consists primarily of mangrove gastropods and dates to approximately 6 750 BP.

Work on the Pilbara coastline (Bradshaw, 1995 and pers. comm.; Clune 2002) provides data from over ten further excavations, demonstrating extensive and broadly based use of marine resources and significantly, the presence of large monospecific (*Anadara granosa*) mounded middens. Excavations of two stratified rockshelters in the same area have revealed marine economic assemblages with a minor terrestrial component. Anadara Shelter returned a basal date of approximately 6 400 BP and the Not-So-Secret Shelter a near basal date of approximately 6 100 BP.

Finally, in the Roebourne area, Veth and O'Brien (1986) reported on excavations of two late Holocene-aged shell middens. The authors argued that the extensive monospecific shell scatters frequently recorded on the Pilbara coastal plain (Abydos Plain) reflect at least a seasonal reliance on marine resources. The apparent disaggregation of middens into non-mounded linear forms was argued to be due to a combination of factors including high residential mobility and the effects of dynamic geomorphic processes on an arid coastal plain. The work of Clune (2002) and Bradshaw (1995) on large mounded middens on the Pilbara coastline, all in protected contexts, lends support to this suggestion. Extensive reworking of stratified middens also has been described from the adjacent arid southwest Kimberley (O'Connor and Veth, 1993:29; O'Connor, 1999b).

We do not think it a coincidence that many areas surveyed within the arid northwest (e.g. O'Connor and Veth, 1993) have a generally sparse hinterland 'signature' yet a plethora of archaeological residues on the coastal fringe. In our view, this patterning does not signify coastal tethering *per se*, but rather is the outcome of different behavioural strategies employed by the same groups; strategies designed to minimise risk and occurring as a fundamental expression of extreme organisational and social fluidity.

Discussion of the Montebello Islands and other regional evidence – with reference to key research questions

The archaeological evidence from Noala and Hayne's Caves, when taken in combination, provide increasing resolution of human responses to the last marine transgression on the now drowned coastal plains of northwest Australia. The key research questions identified at the start of this paper will now be addressed.

1) What does the archaeological record from the Montebello Islands reveal about human use of marine systems at the critical times of 40 ka, 30 ka and < 10 ka when relatively high sea levels brought marine resources close to the sites (cf. Barker, 1999, 2004; Morse, 1999; O'Connor, 1999)?

It is of interest that Noala Cave registers exploitation of coastal resources at approximately 30 000 BP and then again from after *circa* 10 000 BP, but there is no apparent occupation at the earlier time of 40 000 BP when humans are known to be living elsewhere in the northwest region. This is an identical finding to the situation at North West Cape as documented by Morse (1993b) and Przywolnik (2002b, 2005), where the earliest dates all fall into the period 34–30 000 BP. O'Connor and Veth (2000) recently reviewed the wider evidence for Pleistocene exploitation of marine resources in Greater Australia. They concluded that many sites that had the capacity to register earlier coastal occupations, such as Mandu Mandu Creek, Noala Cave, sites on the Vanimo coast of New Guinea, and so on, have all failed to produce evidence of earlier coastal exploitation. Instead, their basal dates all fall into the 30–34 000 bracket, significantly post-dating first occupation of the continent by between 10–30,000 years, depending on whether one wishes to accept a 'short' or a 'long chronology' for the settlement of Sahul (Allen and O'Connell, 1995; Chappell, 2000). According to O'Connor and Veth (2000), this apparently anomalous situation is best explained with reference to a suggestion made originally by Horton (1981); that the degree of orientation towards, or reliance on, coastal resources may well have increased as conditions in the interior deteriorated during the hyperarid phase of the LGM.

A scenario of this kind, where essentially desert-oriented people (see Veth, Smith and Hiscock, 2005) are later impelled to adopt a mixed terrestrial-maritime lifestyle due to excessive aridity of the inland regions, certainly fits comfortably with the evidence from the Montebello Island sites. As documented herein, the Pleistocene faunal assemblages suggest broad-spectrum exploitation of a wide variety of mammals and reptiles, drawn from all locally available terrestrial habitats including dunefields, sand plains and rocky plateaux. Similarly, the lithic assemblages, though small in quantity, point to strong links with the hinterland region, either through large-distance episodic movements or through exchange networks. Whatever their specific history, the human populations that occupied Noala Cave through this period were well versed with survival in an essentially arid environment, irrespective of their degree of proximity to the coast.

In all of the north-western Australian sites, evidence for maritime 'contact' in the Pleistocene is extremely limited. This is particularly true of Noala Cave where very low numbers and weights of marine shellfish were recovered from spits inferred to date to before approximately 10 000 BP. Interestingly enough, each of the two shellfish species found in the lower level of this site are species that can be transported for quite large distances by enclosing them in a clump of mud. Given the reconstructed shoreline scenario based on bathymetric data and known sea level changes, it is assumed that the shellfish from the Pleistocene level of Noala Cave were transported at least 10 km inland (similar distances for Pleistocene shellfish transportation have been obtained recently from excavations in East Timor, Peter Veth, pers. obs) .

The first evidence of more systematic exploitation of coastal resources from Noala Cave dates to around 10 000 BP, at which time the coastline was still probably 10 km or more from the site. At this time marine resources were evidently being used in a complementary fashion, with a maintained strong emphasis on terrestrial resources. This observation notwithstanding, there is little evidence to suggest that marine habitats before sea level stabilisation were necessarily depauperate, at least for the early Holocene period (*contra* Beaton, 1995). This conclusion is in line with the findings from other sites in northern Australia and the southwest Kimberley (O'Connor, 1999b) including North West Cape (Morse, 1999; Przywolnik, 2002b, 2005), and Nara Inlet 1 in the Whitsundays (Barker, 1999, 2004).

By the time the sea had reached the present embayment of Campbell Island, occupation had transferred from Noala to Hayne's Cave. Evidence of exploitation of coastal resources had also increased to the point where a rich, diverse shell midden had begun to accumulate in the Hayne's Cave, albeit still with a strong, complementary

Veth, Aplin, Wallis, Manne, Pulsford, White and Chappell

terrestrial focus. This pattern of balanced resource use was probably sustained for at least 500 years, during which time the coastal plain habitats were slowly but inexorably swallowed by the rising sea. The uppermost three spits of Hayne's Cave probably date to the time of final insularisation. The diversity and abundance of terrestrial fauna decline in those levels, despite the fact that most classes of marine fauna continue without obvious reduction in quality or quantity.

From this analysis of the archaeological record, we feel confident in concluding that marine resources were exploited by people utilising the Montebello Island caves whenever the coast came to within an economically viable distance of the sites. This is consistent with the conclusions reached by Morse, O'Connor and others based on excavations in adjacent parts of north-western Australia.

2) What insights do the cave sequences provide regarding Pleistocene and early Holocene material culture?

Stone artefacts alone were recovered from Noala Cave, while bone and shell artefacts were additionally recovered from Hayne's Cave. In both sites, the artefact assemblage is limited in quantity, although the stone artefacts are manufactured from comparatively diverse raw materials.

Artefacts made on silicified calcrete are present in varying yet low numbers throughout both sites; this material is presumed to come from local sources on nearby Barrow Island or from equivalent drowned contexts. The technology of manufacture of these assemblages, which mainly comprise *debitage*, appears opportunistic, and probably reflects the low quality and ready availability of this raw material (see detailed discussion in Appendix 2). Minimal retouch was only recorded on several pieces.

The lithic assemblages also include a range of 'exotic' materials derived from distant mainland sources, perhaps as far as 120 km or more to the south-east. Several cores from a basic volcanic rock, including a large discoidally flaked single platform 'horsehoof' core, provide evidence for more carefully controlled reduction.

Artefacts made from exotic raw materials are noticeably absent from the upper five spits of Hayne's Cave. These spits are assumed to date to the final phase of transgression, during which the land bridge to the contemporary Pilbara coastline was lost and replaced by a gradually widening water barrier. Given the relatively small size of the islands and limited availability of freshwater, we consider that the Montebello Islands are unlikely to have supported resident human populations after full insularisation. However, it is possible

that the islands were visited for some time after initial cut-off (perhaps initially from Barrow Island), until such time that it became too hazardous for the perceived returns.

The presence of three possible shell 'beads' and a bone point in Hayne's Cave is consistent with similar finds made by Morse (1993b) from the site of Mandu Mandu Creek at North West Cape and Balme (2000; Balme and Morse, 2004) at the Riwi rockshelter site in the inland south Kimberley. Przywolnik (2002b:203-205) also recovered one complete and three bone point tips made on macropod long-bones from the Holocene levels of the C99 site. These sparse finds give a limited but tantalizing glimpse into more perishable components of the material culture.

3) What do the Montebello Island sites reveal about human responses to post-glacial sea level rise? In particular, is there evidence for a time lag between sea level rise and midden development, as suggested by Beaton (1995), or is the appearance of middens archaeologically instantaneous?

Despite the refinement of local bathymetric data for the Montebello Islands and improvements in general sea level modelling (Lambeck and Chappell, 2001), it remains difficult to control for locality-specific sediment accumulation in the channels which now separate the islands, for the effects of local isostatic responses and indeed the normal margins of error which might pertain to C14 dates on shellfish and estimation of the Oceanic Reservoir Correction factor for these northern tropical waters. Nevertheless time-lags in the order of 1 000 years or more between approximation of the coast and midden deposition should be discernable. As discussed at length by Veth (1993, 1995, 1999), comparisons of the expected position of the shoreline and the dates for midden accumulations (after appropriate corrections) fail to support the notion of a major time lag. Instead, there is evidence for exploitation of marine resources in the sites as soon as the transgressive shoreline is estimated to be less than 10 to 5 km from the sites. The densest midden deposits occur in Hayne's Cave, dating from approximately 7 800 to 7 200 BP, immediately following the arrival of the coastline near its present location. The sequence and timing of changes in occupation patterns are thus broadly consistent with expectations based on general sea level modelling and indicate that marine resources were exploited at the sites whenever they were within reasonable travelling distance.

As outlined earlier, a similar spread of dates for exploitation of marine resources has been obtained from middens along the north-west coast of Australia. From this we conclude that the rapid response to marine transgression recorded so precisely in the Montebello archaeological record is a true reflection of a more general, regional pattern.

4) To what extent were the transgressing shorelines depauperate in resources in comparison with habitats and fauna so efficiently exploited by coastal foragers in the recent past?

Bowdler (1999:82) specifically posed the question of when the mangrove/estuarine adaptation witnessed during the mid- to late Holocene along much of the central and northern coastline of Western Australia emerged. Mangrove communities were an important traditional source of animal foods including shellfish, fish, crustaceans and even vertebrates, as well as certain plant foods, and their presence or otherwise along a coastline can strongly influence its suitability for human occupation.

Significant insights into this issue have already come from various coastal excavations spanning the terminal Pleistocene to mid-Holocene. To begin with, excavations on the Burrup Peninsula, Pilbara coastline and the Dampier Archipelago have established that middens formed soon after the sea came within close proximity of the contemporary coastline. These middens contain faunas that are specifically associated with mangrove communities (cf. Bradshaw, 1995; Clune 2002; Veth, 1999). Furthermore, Elizabeth Bradshaw (pers. comm.) has obtained dates as early as approximately 8 000 BP for mangrove shellfish on islands of the Dampier Archipelago where the offshore profile is steep and evidence of earlier coastal occupations might be expected. Finally, there are sites on the present coastline that date to within the period 8 000 to 6 000 BP and which contain abundant remains of mangrove shellfish such as *Terebralia* sp. Thus there is little doubt that mangrove communities were at least locally present along the transgressing northwest coastline from at least 8 000 BP.

The location of the Montebello Islands allows us to push the record for systematic exploitation of coastal resources on the plain back to over 10 000 BP. They provide firm evidence for the presence at that time of mangrove communities and the exploitation of diverse fauna from this habitat. As Veth (1999:67) noted:

> ...the presence of mangroves is confirmed by pollen in cores taken from the north of the Sahul shelf (van der Kaars 1991) with a relative date of 12,000 BP. Firm evidence for mangrove communities comes from the Montebello Islands with the continuous presence of the gastropods *Terebralia* sp. and *Telescopium* sp., mud crab (*Scylla serrata*) and mud lobster

(*Thalassinia anomola*), all dating to between 10,000 BP and 7,500 BP

When the Montebello sequences are combined with those of the Dampier Archipelago, the Abydos Plain and the sites from North West Cape (Morse, 1999; Przywolnik, 2002b, 2005), it becomes clear that mangrove communities were not only present through the entire post-glacial transgressive episode, but also that they were quite possibly every bit as diverse as those found in the area today. Admittedly, less is known about the character of the coastline when it lay on the margin of the continental shelf. However, based on the evidence at hand, it seems likely that mangrove communities were present at least discontinuously along the edge of the continental shelf and that they were not entirely depauperate in resources.

Veth (1999:67) has described a broad regional trend in which an early dominance of mangrove gastropods, such as *Terebralia palustris* and *Turbo cinereus*, gives way by *circa* 3 600 BP to ubiquitous assemblages of the bivalve *Anadara granosa*, with some regional variation in the precise timing of this change. Veth (1999:67), Bowdler (1999) and O'Connor (1999) have all noted the widespread mid- to late Holocene retraction of mangrove communities along extensive sections of the arid western coastline. This may in fact be an even more widespread phenomenon, following the so-called 'big swamp' phase of mangrove expansion in the early to mid-Holocene (Woodroffe, Chappell, Thom and Wallensky, 1989).

5) What do the early dates for coastal exploitation contribute to the emerging debate about environmentally vs socially driven processes of cultural change

This complex issue cannot be fully addressed in this context (see also Veth *et al.*, 2000). However, we consider it relevant to further emphasise the fact that the changing pattern of procurement and processing of both marine and terrestrial resources from the Montebello Islands cave sites closely mirrors the reconstructed environmental changes in the surrounding area. Importantly, the changes observed between the upper portions of Noala Cave and the lower part of the Hayne's Cave deposit suggest that the exploitation of marine resources from these adjacent sites intensifies at precisely the time when general sea level models suggest the coast becomes proximal, and considerably before the time that general models for social intensification are generally evoked for coastal assemblages (cf. Hall and McNiven, 1999a). Furthermore, the nature of extractive activities conducted from Hayne's Cave shows subtle changes in response to the progressive loss of terrestrial habitats, suggesting a staged process of abandonment of the site. This apparent close link between resource availability and the spatial patterning and intensity of various extractive activities urges caution in overly simplistic interpretation of other regional sequences, especially where the environmental boundary conditions are less clear cut.

It is worth commenting in this context on the large monospecific mounds that are present on the adjacent Pilbara coastline. These appear to date from after 4 500 BP (uncorrected for ORE) (e.g. Clune, 2002) but their significance is moot. Veitch (1999) has urged us to consider the link between the emergence of monospecific mounds in northern Australia and behavioural factors, such as a shift towards *r*-selected species. However, O'Connor (1999) has argued that the emergence of mounds reflects ecological changes in molluscan communities and patterns of exploitation in response to shifts in the northwest monsoon and local sedimentary conditions. Interestingly enough, the earliest dates for mounds show a cline from the Pilbara, through the Kimberley coast, appearing latest in Arnhem Land (O'Connor, 1999; Figure 7).

6) Is there evidence for a residential population by the time the island group has formed and for how long does it persist?

The history of exploitation and abandonment of this small island group presents a familiar biogeographic scenario (*cf.* Jones, 1977; O'Connor, 1992; Sim, 1998, 2002, 2004). The earliest date for occupation of Hayne's Cave (with ORE applied) is approximately 7 800 BP while the most recent date is approximately 7 000 BP. Through most of the period bracketed by these dates it is likely that the Montebello Islands were connected to Barrow Island and portions of the mainland via an exposed coastal plain. However, as argued in a previous section, the last phase of occupation of Hayne's Cave differs in character from the earlier periods and is thought to narrowly post-date severance of one or more of these land bridges. Evidence for this interpretation includes the absence of exotic lithics from the five most recent spits, a reduction in the quantity and variety of terrestrial fauna in these same spits, and gradual shifts in the chemistry of the sediments suggesting an increasingly maritime influence. For perhaps a century or more a combined Barrow and Montebello Island landmass would have been narrowly separated from the mainland by the nascent Mary Anne Passage, with final fragmentation into discrete island groups occurring not long afterwards (see Figure 2). For a while, these water barriers may have been crossed by people using watercraft or other means. However, soon after 7 000 BP the islands would have become remote both from each other and

even more importantly, from the mainland. The islands were almost certainly abandoned at this time, with no evidence found either on the Montebello Islands or on Barrow that would suggest any mid- or late Holocene occupation. While the marine systems of the numerous small islands of the Montebello group clearly remained very productive through to the present day (*cf.* Morris, 1991), they appear to have been nonviable on their own as territories or homelands for groups of hunter-gatherers. This is in contrast to the situation in the Sir Edward Pellew Group of islands in the Gulf of Carpentaria, where Sim (2002, 2004) has shown that despite abandonment soon after insularisation, Vanderlin Island was apparently reoccupied from *circa* 4 000 BP. Similarly, High Cliffy Island lying adjacent to the west Kimberley coastline has evidence for Aboriginal use of the island at 6 700 bp, post-dating its separation from the mainland, as well as in the mid- to late Holocene period, suggesting it remained part of the territory of a mainland group (O'Connor, 1994, 1999a).

7) Does the Montebello archaeological record yield any information of a general nature regarding the human ecology of Pleistocene foragers on an arid coastline, particularly in regard to how they would have structured their use of littoral and hinterland resources.

We believe that the earliest foragers of the arid coastal plains of northwest Australia would have had much in common with contemporaneous desert occupants; and further, that the greater reliability and concentration of marine resources would not have overshadowed a fundamental similarity in their extractive and socioeconomic systems. Notwithstanding this suggestion, we accept the proposition of O'Connor and Veth (1993; see also Veth, 1995, 1999) that the coastal-oriented populations probably displayed a distinctive admixture of desert and coastal adaptations. This in turn may have allowed such groups to enjoy a broader range in their residential mobility patterns, with periodic aggregations quite possibly focused on coastal sites. It is further plausible that these sites may have been especially important during normal 'dry' seasons and in the context of episodic droughts (see discussions also in Clune 2002).

In a recent paper, Veth (2005: Table 6.1) outlined some of the behavioural correlates and likely archaeological signatures that might be predicted for arid zone groups utilising a landscape characterised by an unpredictable and scarce resource structure. Behavioural correlates include large territories with permeable boundaries, the use of very high and opportunistic residential mobility patterns, high dietary breadth and high levels of information exchange. The introduction of relatively predictable and abundant marine resources into this scenario serves to shift some of the expectations of the 'pure desert' model but possibly for only part of the normal foraging cycle. This raises the likelihood that groups may have switched mobility patterns on a regular basis and that the use of habitation sites on the coast, presumably located near freshwater sources, would have been more logistically organised (after Binford, 1980). In other words, greater intensity and permanency of occupation during the coastal phase of occupation might be followed by a switch in strategy toward wide dispersion and systematic exploitation of vast portions of the interior. While this scenario is not unique to north-west Australia, the scale of the territory exploited and the strength of the contrast between coastal and interior land use patterns quite possibly is.

Conclusion

This monograph is a synthesis of the first phase of archaeological analysis of two stratified cave sites on Campbell Island in the Montebello Group of northwest Australia. These sites clearly have great potential to directly register human adaptations on and near an arid coastline during a time of extraordinary changes in the palaeogeography of northwest Australia—the terminal Pleistocene and early Holocene—when a vast northwest arid coastal plain with its low, rocky plateaux was drowned to produce the far flung island groups of today. Rich faunal assemblages provide unique insights into the terrestrial and marine landscapes for several key intervals during the period 30 000 – 7 000 BP as the local environmental context changed in response to the post-glacial marine transgression.

The Montebello Island sites add significantly to the evidence for longterm use of marine resources in the arid northwest and give a particularly detailed record of the fascinating period spanning the final millennia of the terminal Pleistocene and the early Holocene. Sporadic, low level occupation of the Montebello caves occurred between approximately 30 000 and 10 000 BP, paralleling the record from similar-aged stratified sites at North West Cape (Morse, 1996, 1999; Przywolnik, 2002b, 2005). More intensive habitation involving utilisation of both marine and terrestrial resources is registered from approximately 10 000 BP, with a further intensification over the period 7 800–7 000 BP as the sea approached its current position. This evidence adds significant new detail to an emerging picture of regional exploitation of marine resources during the early and mid-Holocene, as revealed by excavations on the Pilbara coast, the Burrup Peninsula and the Dampier Archipelago (cf. Bradshaw, 1995; Morse, 1996; Veth 1999; Clune, 2002; Przywolnik, 2002b, 2005). When these early to mid-Holocene sequences are considered in the context of the

changing palaeogeography of the continental shelf, it must be concluded that systematic exploitation of marine resources has had an extremely lengthy and near-continuous history on this arid coastal plain.

Some of the earlier stone artefacts recovered from both Noala and Hayne's Caves have a mainland provenance. This implies that groups either had large territories extending over 100 km into the interior or else had exchange relations with groups who were largely resident in the interior. This evidence for procurement of hinterland resources is only lost during the final phase of occupation of Hayne's Cave. We infer that, at this time, the greater Barrow and Montebello Island landmass was probably already cut off by a narrow water gap but perhaps not so extensive as to prohibit movement to and from the mainland.

The mammalian fauna in the Montebello cave deposits are highly diverse and significantly richer than those currently found in existence there or on nearby Barrow Island. The diversity of medium-sized mammals is also higher than in the contemporaneous Cape Range faunas, owing to the presence of a number of sand plain specialists that must have occupied the now-submerged coastal plains with its extended dunefields. Faunal evidence from Hayne's Cave suggests that the coastal plain habitat went through a brief phase of positive water balance, allowing herbaceous plant growth and the invasion from the subtropical Kimberley region of at least one mammal species – the northern nail-tail wallaby, *Onychogalea unguifera*. This brief 'climatic optimum' was probably due to the combination of regional temperature rise with flooding of the outer portion of the shallow, broad continental shelf, providing ideal conditions for cloud build-up and possibly for local precipitation (van der Kaars and De Deckker, 2002).

Human exploitation of the continental shelf was complemented by littoral resources when the coastline became proximal to the sites, by approximately 10 000 BP. A marked increase in the exploitation of marine resources is registered in Noala Cave at approximately 10 000 BP, when shell starts to accumulate in some abundance. Earlier occurrences of shellfish are few in number and limited to species that might be easily transported in clumps of mud or species whose use may not have been dietary. The dense midden of Hayne's Cave, dated to between approximately 7 800 –7 000 BP, accumulates at precisely the time when the coast is estimated to be within a short distance of the sites. The final phase of occupation of Hayne's Cave is characterised by the loss of most terrestrial fauna coupled with the continued use of marine resources. The dating and content of these midden layers leads us to conclude that site usage adopted a marine focus as soon as the coast came within a reasonable foraging range from the site (*cf.* Bird, 1996).

There is no evidence from these two sites for any appreciable time-lag between marine transgression and the availability and, more importantly, the procurement of marine resources (*contra* Beaton, 1995). To the contrary, marine fauna recovered from both Noala and Hayne's Caves illustrate that the full range of marine habitats which were available to coastal foragers during historic times, were also present throughout the transgressive phase. Among the shellfish species recovered are denizens of mature reef flats, rocky foreshore substrates, intertidal mudflats and, importantly, mangrove communities. The occurrence of mangrove dwelling taxa is especially important given Bowdler's (1999:82) voicing of doubts regarding the presence of mangroves along the edge of the exposed continental shelf and their ability to recolonise during times of sea level rise. The presence of mangrove shellfish and crab from approximately 12 000–7 000 BP provides firm evidence of a mangrove community keeping pace with sea level rise and a transgressive coastline. Importantly, there is also little reason to believe that the littoral environment was depauperate either during a transgressive phase or during the period when sea level stabilised at the current high stand.

The final phase of insularisation and eventual isolation of the sites from the mainland is well represented in the economic assemblages of Hayne's Cave. Terrestrial faunal elements and mainland lithic materials are largely absent from the upper spits of the site and occupation ceases at approximately the time that the islands would have become the far flung outliers of today. There is no evidence from these sites, or from any other of the (minimal) cultural materials located on the islands, for the existence of either a residential group or seasonal usage after 7 000 BP. The islands were evidently abandoned at this time and never reoccupied. If they were ever visited again prior to the historic period, then any such events left no appreciable archaeological record.

In summary, the dating and archaeological context of the Montebello sites appears to closely mirror the environmental events associated with sea level rise, the increasing proximity of marine resources and the eventual loss of the arid coastal sand plains. While territorial and mobility adjustments may have been made during the LGM, resulting in increased focus on the Pleistocene coastline, there seems little need to invoke social processes to explain the structure or timing of occupation of these sites.

We conclude by reflecting on the possibly unique characteristics of foragers on an arid coastline. While the fundamental elements of a 'desert adaptation' are likely to have obtained for much of the economic cycle, such as a generally high residential mobility pattern and low population numbers, we predict that a switch in strategy occurred when groups resided by the sea. The relative abundance of high protein and fat resources, and the presumed shift towards a more logistical strategy, might well result in more intensive occupation at habitation sites, resulting in higher archaeological visibility overall. Periodic coastal aggregations may well have served the same function as those of the interior, eventually becoming centres of production for intensifying social and economic systems.

References

Akerman, K. (1973). Aboriginal baler shell objects in Western Australia. *Mankind* **9** (2):124-125.

Allen, G. (1997). *Marine Fishes of Tropical Australia and South-East Asia.* Western Australian Museum, Perth.

Allen, J. and O'Connell, J.F. (eds) (1995). *Pleistocene to Holocene in Australia and Papua New Guinea. Antiquity* **69**.

Aplin, K.P. and Donnellan, S.C. 1999). An extended description of the Pilbara Death Adder, *Acanthophis wellsi* Hoser (Serpentes: Elapidae), with notes on the Desert Death Adder, *A. pyrrhus* Boulenger, and identification of a possible hybrid zone. *Records of the Western Australian Museum* **19**: 277–298.

Aplin, K.P., Adams, M. and Cowan, M.A. (*in press*). Systematics and biogeography of the herpetofauna of the Carnarvon Basin region of Western Australia. *Records of the Western Australian Museum, Supplement.*

Aplin, K., Baynes, A., Chappell, J. and Pillans, B. (2001). Pliocene and Quaternary vertebrate faunas from a succession of karstic and related coastal deposits on Barrow Island, northwestern Australia. Abstract of a paper presented at the 2001 Biennial Conference of the Australian Quaternary Association., Port Fairy, February 2001.

Archer, M. (1977). *In*, Gould, R.A. (ed.), *Puntutjarpa Rockshelter and the Australian Desert Culture. Anthropological Papers of the American Museum of Natural History*, Number **54(1)**.

Armstrong, K.N. and Ainstee, S.D. (2000). The ghost bat in the Pilbara: 100 years on. *Australian Mammalogy*, **22**: 93-100.

Balme, J. (2000). Excavations revealing 40,000 years of occupation at Mimbi Caves, south central Kimberley, Western Australia. *Australian Archaeology* **50**:1–5.

Balme, J. and Morse, K. (2004). To bead or not to bead. Unpublished paper presented at the Australian Archaeological Association Annual Conference, December 12-15 2004, Armidale.

Barker, B. (1999). Coastal occupation in the Holocene: environment, resource use and resource continuity. pp. 119–128. *In*, Hall, J. and McNiven, I.J. (eds), *Australian Coastal Archaeology. Research Papers in ANH* Number **31**. ANH Publications, Research School of Pacific and Asian Studies, The Australian National University, Canberra.

Barker, B. (2004). *The Sea People: Late Holocene maritime specialisation in the Whitsunday Islands, central Queensland. Terra Australis* Number 20. Pandanus Press, The Australian National University, Canberra.

Bates, D. (n.d.). *The Tribal Organisation and Geographical Distribution: Western Australia.* Manuscript number 92, Australian Institute of Aboriginal and Torres Strait Islander Studies, Canberra.

Baynes, A. (2000). Original mammal faunas of the Carnarvon Basin, based on fossil material from the surfaces of small caves. Appendix 2. Pp. 479-510. *In*, McKenzie, N.L., Hall, N.J. and Muir, W.P. Non-volant mammals of the southern Carnarvon Basin, Western Australia. *Records of the Western Australian Museum, Supplement* No. 61.

Baynes, A. and Johnson, K.A. (1996). Chapter 14.The contributions of the Horn Expedition and cave deposits to knowledge of the original mammal fauna of central Australia. Pp. 168-186. *In*, Morton, S.R. and Mulvaney, D.J. (eds) *Exploring Central Australia*: Society, the environment and the 1894 Horn Expedition. Surrey Beatty and Sons Pty Ltd, Chipping Norton.

Baynes, A. and Jones, B. 1993). The mammals of Cape Range Peninsula, Western Australia. *Records of the Western Australian Museum* Supplement Number **45**:207–225.

Baynes, A., Merrilees, D. and Porter, J.K. (1975). Mammal remains from the upper levels of a late Pleistocene deposit in Devil's Lair, Western Australia. *Journal of the Royal Society of Western Australia* **58**(4):97-117.

Beard, J.S. (1975). *The Vegetation of the Pilbara Area. Vegetation Survey of Western Australia. 1:1,000,000 Vegetation series. Explanatory Notes to Sheet 5.* Pilbara. University of Western Australia Press, Nedlands.

Beaton, J.M. 1995). The transition on the coastal fringe of greater Australia. *Antiquity* **69**: 798–806.

Berry, P.F. (1993). *Survey of the marine fauna and habitats of the Montebello Islands, August 1993.* Unpublished report to the Western Australian Museum, Perth.

Binford, L.R. 1980). Willow smoke and dogs' tails: Hunter-gatherer settlement and archaeological site formation. *American Antiquity* **45**: 4–20.

Bird, D.W. 1996). *Intertidal Foraging Strategies Among the Meriam of the Torres Strait Islands, Australia: An evolutionary ecological approach to the ethnoarchaeology of tropical marine subsistence.* Unpublished PhD thesis, Department of Anthropology, University of California, Berkeley.

Bird, M. (1992). The impact of tropical cyclones on the archaeological record: An Australian example. *Archaeology in Oceania* **27**:75–86.

Bowdler, S. (1983). Sieving seashells: Midden analysis in Australian archaeology. pp. 135–144. *In*, Connah, G. (ed.), *Australian Field Archaeology: A guide to techniques.* Australian Institute of Aboriginal Studies, Canberra.

Bowdler, S. 1984). *Hunter Hill, Hunter Island: Archaeological investigations of a prehistoric Tasmanian site.* *Terra Australis* Number **8.** Department of Prehistory, Research School of Pacific Studies, The Australian National University, Canberra.

Bowdler, S. (1999). Research at Shark Bay, WA, and the nature of coastal adaptations in Australia. pp. 79-90. *In*, Hall, J. and McNiven, I. (eds), *Australian Coastal Archaeology.* Department of Archaeology and Natural History, Research School of Pacific and Asian Studies, The Australian National University, Canberra. *Research Papers in Archaeology* Number 31.

Bradley, A.J., Kemper, C.M., Kitchener, D.J., Humphreys, W.F., How, R.A. and Schmitt, L.H. (1988). Population ecology and physiology of the common rock rat, *Zyzomys argurus* (Rodentia: Muridae) in tropical northwestern Australia. *Journal of Mammology* **69**: 749–764.

Bradshaw, E. 1995). Dates from archaeological excavations on the Pilbara coastline. *Australian Archaeology* **41**: 37–38.

Braithwaite, R.W. and Begg, R.J. (1998). Northern Quoll. pp. 65-66. *In*, Strahan, R. (ed.), *Mammals of Australia* (2nd ed.). Reed Books, Chatswood.

Burbidge, A.A. 1971). *The Significance of the Montebello Islands. Department of Fisheries and Fauna Report* Number **9**. Department of Fisheries and Fauna, Perth.

Bureau of Meteorology. 1998). <http://www. bom.gov.au/climate/averages/tables/cw_005058.sh tml; accessed 20 June 2004>

CALM——see Conservation and Land Management

Casteel, R.W. (1976). *Fish Remains in Archaeology and Palaeo-environmental Studies.* Academic Press, London.

Cavalche, M.L. and Pulido-Bosch, A. 1994). Modelling the effects of saltwater intrusion dynamics for a coastal karstified block connected to a detrital aquifer. *Groundwater* **32(5)**: 767–777.

Chaloupka, G. (1993). *Journey in Time.* Reed Books, Sydney.

Chappell, A. (1994). *Mineralogical and Chemical Characteristics of Sediments within Caves on the Montebello Islands of Western Australia and their Association with Sealevel and Climate Change.* Unpublished report held by Woodside Offshore Petroleum, Perth.

Chappell, J. 1982). *Some Effects of Cyclonic Waves and Tidal Currents of the Coast and Continental Shelf near Dampier, Western Australia.* Unpublished report to Woodside Australian Energy.

Chappell, J. (1994). Upper Quaternary sea levels, coral terraces, oxygen isotopes and deep-sea temperatures. *Journal of Geography* **103**: 828-840.

Chappell, J. (2000). Pleistocene seedbeds of western Pacific maritime cultures and the importance of chronology. *In*, O'Connor, S. and Veth, P. (eds), *East of Wallace's Line: Studies of past and present maritime cultures of the Indo-Pacific region.* Balkema, Rotterdam. *Modern Quaternary Research in Southeast Asia* **16**: 77–98.

Chappell, J. and Thom, B.G. 1977). Sea levels and coasts. pp. 275-291. *In*, Allen, J., Golson, J. and Jones, R. (eds), *Sunda and Sahul: Prehistoric Studies in Southeast Asia, Melanesia and Australia.* Academic Press, London.

Chappell, J. and Thom, B.G. 1986). Coastal morphodynamics in north Australia: review and prospect. *Australian Geographical Studies* **24**: 110–27.

Clune, G. (2002). Abydos: An archaeological investigation of adaptations on the Pilbara coast, northwest Western Australia. Unpublished PhD thesis, Centre for Archaeology, University of Western Australia, Perth.

Coleman, N. (1975). *What Shell is That?* Paul Hamlyn, Dee Why West.

Colley, S.M. (1986). Site formation and archaeological fish remains: An ethnohistorical example from the Northern Isles, Scotland. pp. 34–41. *In*, Brinkhuizen, D.C. and Clason, A.T. (eds), *Fish and Archaeology: Studies in osteology, taphonomy, seasonality and fishing methods.*

British Archaeological Report International Series Number **294**, Oxford.

Collins, L.B., Zhu, Z.R., Wyrwoll, K.H., Hatcher, B.G., Playford, P.E., Chen, J.H., Eisenhauer, A. and Wasserburg, G.J. (1993). Late Quaternary evolution of coral reefs on a cool water carbonate margin: the Abrolhos Carbonate Platforms, southwest Australia. *Marine Geology* **110**: 203–212.

Conservation and Land Management. (2005). *The Montebello-Barrow Islands Marine Conservation Reserves.* <http://www.calm.wa.gov.au/national_parks/marine/montebello-barrow/index.html; accessed 15 May 2005>

Cooper, N.K., Aplin, K.P. and Adams, M. (2000). A new species of false antechinus (Marsupialia: Dasyuromorphia: Dasyuridae) from the Pilbara region, Western Australia. *Records of the Western Australian Museum* **20**: 115-136.

Corbett, L.K. (1995). *The Dingo in Australia and Asia.* New South Wales University Press, Sydney.

Crawford, I. 1986). *Report and recommendations from a visit to the Montebello Islands, August 1986.* Unpublished Western Australian Museum report to the Department of Conservation and Environment.

Cronin, L. (2001). *Australian Reptiles and Amphibians.* Envirobook, Annandale.

David, B. and Stanisic, J. (1991). Land snails in Australian archaeology: Initial results from Echidna's Rest (north Queensland). *The Artefact* **14**:19–24.

Davis, R.A. and Clifton, H.E. (1987). Seal-level change and the preservation potential of wave dominated and tide dominated coastal sequences. pp. 167–178. *In*, Nummedal, D., Pilkey, O.H. and Howard, J.D. (eds), *Seal-Level Fluctuation and Coastal Evolution.* Society of Economic Palaeontologists and Mineralogists, Tulsa.

Dickman, C.R., Daly, S.E.J. and Connell, G.W. (1991). Dietary relationships of the Barn Owl and Australian Kestrel on islands off the coast of Western Australia. *Emu* **91**: 67–92.

Dortch, C. and Morse, K. 1984). Prehistoric stone artefacts on some offshore islands in Western Australia. *Australian Archaeology* **19**: 31–47.

Douglas, A.M. (1967). The natural history of the Ghost Bat, *Macroderma gigas* (Microchiroptera, Megadermatidae), in Western Australia. *The Western Australian Naturalist*, **10**: 125-137.

Frank, E.F., Mylroie, J., Troester, J., Alexander Jr, E.C. and Carew, J.L. (1998). Karst development and speleogenesis Isla de Mona, Puerto Rico. *Journal of Cave and Karst Studies* **60(2)**: 73–83.

Friend, J.A. (1990). Status of bandicoots in Western Australia. pp. 73–84. *In*, Seebeck, J.H., Brown, P.R., Wallis, R.L. and Kemper, C.M. (eds), *Bandicoots and Bilbies.* Surrey Beatty in association with The Australian Mammal Society, Sydney.

Gould, R. A. (1977) *Puntutjarpa Rockshelter and the Australian desert culture.* Anthropological papers of the AMNH, American Museum of Natural History, New York.

Gould, R.A. 1996). Faunal Reduction at Puntutjarpa rockshelter, Warburton Ranges, Western Australia. *Archaeology in Oceania* **31**: 72–86.

Grayson, D.K. (1984). *Quantitative Zooarchaeology: Topics in the analysis of archaeological faunas.* Academic Press, Orlando.

Hall, J. and McNiven, I. J. (eds) (1999a). *Australian Coastal Archaeology.* ANH Publications, Research School of Pacific and Asian Studies, The Australian National University, Canberra. *Research Papers in ANH* Number **31**.

Hall, J. and McNiven, I.J. (1999b). Australian coastal archaeology: Introduction. pp. 1–5. *In*, Hall, J. and McNiven, I.J. (eds), *Australian Coastal Archaeology.* ANH Publications, Research School of Pacific and Asian Studies, The Australian National University, Canberra. *Research Papers in ANH* Number **31**.

Hall, L.S. (1998). Black Flying-fox. pp. 432-433. *In*, Strahan, R. (ed.), *The Mammals of Australia* (2nd ed.). Reed Books, Chatswood.

Hedges, E.M. and Millard, A.R. (1995). Bones and groundwater: towards modelling of diagenetic processes. *Journal of Archaeological Science* **22**: 155–164.

Hill, B.J. (1974). Salinity and temperature tolerance of zoeae of the portunid crab *Scylla serrata*. *Marine Biology* **25**:21–24.

Hill, F.L. (1955). Notes on the natural history of the Monte Bello Islands. *Proceedings of the Linnean Society of London* **165(2)**: 113–124.

Hocking, R.M., Moors, H.T. and van der Graff, W.J.F. 1987). *Geology of the Carnarvon Basin, Western Australia. Western Australian Geological Survey Bulletin Number 133.*

Hook, F., McDonald, E., Paterson, A.G., Souter, C. and Veitch, B. (2004). *Cultural Heritage Assessment and Management Plan - Proposed Gorgon Development, Pilbara, Northwestern Australia*. Environmental Resources Management Australia Ptd Ltd and Gorgon Australian Gas, Perth.

Hope, J. (1983). Recovery and analysis of bone in Australian archaeological sites. pp. 126–134. *In*, Connah, G. (ed.), *Australian Field Archaeology: A guide to techniques*. Australian Institute of Aboriginal Studies, Canberra.

Hopley, D. (1984). The Holocene 'High Energy Window' on the Central Great Barrier Reef. pp. 151–178. *In*, Thom, B.G. (ed.), *Coastal Geomorphology in Australia*. Academic Press Australian, North Ryde.

Horton, D.R. (1981). Water and Woodland: The peopling of Australia. *Australian Institute of Aboriginal Studies Newsletter* **16**: 21–27.

Horton, D.R. (1994). *The Encyclopaedia of Aboriginal Australia*. Aboriginal Studies Press, Canberra.

Houbrick, R.S. (1991). Systematic review and functional morphology of the mangrove snails *Terebralia* and *Telescopium* (Potomidiidae: Prosobranchia). *Malacologia* **33**(1-2): 289–338.

Humphreys, W.F. 1993). The significance of the subterranean fauna in biogeographical reconstruction: examples from Cape Range Peninsula, Western Australia. *Records of the Western Australia Museum* Supplement Number **45**: 165–192.

Ingleby, S. and Gordon, G. (1998). Northern Nailtail Wallaby. pp. 361-362. *In*, Strahan, R. (ed.), *The Mammals of Australia* (2nd ed.). Reed Books, Chatswood.

Ingleby, S., Westoby, M., and Latz, P.K. (1989). Habitat requirements of the Northern Nail Tail Wallaby *Onychogalea unguifera* (Marsupialia: Macropodidae) in the Northern Territory and Western Australia. pp. 767–782. *In*, Grigg, G., Jurman, P. and Hume, I. (eds), *Kangaroos, Wallabies and Rat-Kangaroos*. Surrey Beatty and Sons, Sydney.

Johannes, R.E. (1991). *Traditional Fishing in the Torres Strait Islands*. CSIRO Division of Fisheries, Hobart.

Johnstone, R.E. (1990). Mangroves and mangrove birds of Western Australia. *Western Australian Museum Supplement* **32**.

Jones, R. (1977). Man as an element of a continental fauna: The case of the sundering of the Bassian Bridge. pp. 317-386. *In*, Allen, J., Golson, J. and Jones, R. (eds), *Sunda and Sahul: Prehistoric Studies in Southeast Asia, Melanesia and Australia*. Academic Press, London.

Kendrick, G.W. and Morse, K. 1982). An Aboriginal shell midden deposit from the Warroora Coast, northwestern Australia. *Australian Archaeology* **14**: 6–12.

Kendrick, P.G. (1993). Biogeography of the vertebrates of the Cape Range peninsula, Western Australia. pp. 193-206. *In*, Humphreys, W.F. (ed.), *The Biogeography of Cape Range, Western Australia. Records of the Western Australian Museum, Supplement* No. 45.

Kershaw, A.P. and Nanson, G.C. 1993). The last full glacial cycle in the Australian region. *Global and Planetary Change* **7**: 1–9.

Lambeck, K. and Chappell, J. (2001). Sea level change through the last glacial cycle. *Science* **292**: 679-686.

Leach, F. and Davidson, J. (2000). Fishing: A neglected aspect of Oceanic economy. pp. 412-426. *In*, Anderson, A. and Murray, T. (eds), *Australian Archaeologist: Collected papers in honour of Jim Allen*. Pandanus Press, The Australian National University, Canberra.

Leatherman, S.P. (1990). Modelling shore response to sea level rise on sedimentary coasts. *Progress in Physical Geography* **14**(4): 447–464.

Limpus, C.J. and Miller, J.D. (1993). Family Cheloniidae. pp. 133–138. *In*, Glasby, C.J., Ross, G.J.B., and Beesley, P.L. (eds), *Fauna of Australia (volume 2A): Amphibia and Reptilia*. Australian Government Publishing Service, Canberra

Lorblanchet, M. (1977). Summary report of fieldwork, Dampier, WA. *AIAS Newsletter* **7**:36–40.

Lorblanchet, M. (1992). The rock engravings of Gum Tree Valley and Skew Valley, Dampier, Western Australia: Chronology and function of sites. pp. 39-59. *In*, McDonald, J. and Haskovec, I.P. (eds), *State of the Art: Regional rock art studies in Australia and Melanesia. Occasional AURA Publication* No. **6**. Australian Rock Art Research Association, Melbourne.

Lyman, R.L. (1994). *Vertebrate Taphonomy*. Cambridge University Press, Cambridge.

Manne, T.H. 1999). *Cultural Responses to the Flandrian Transgression on the Montebello Islands, Northwest Australia*. Unpublished BSc (Hons) Thesis, Department of Archaeology and

Anthropology, James Cook University, Townsville.

Marshall, B. and Cosgrove, R. (1990). Tasmanian Devil (*Sarcophilus harisii*) scat-bone: signature criteria and archaeological implications. *Archaeology in Oceania* **25**: 102–113

Martinez, J.I., De Deckker, P. and Barrows, T.T. (1999). Palaeoceanography of the last glacial maximum in the eastern Indian Ocean: planktonic and foraminiferal evidence. *Paleogeography, Paleoclimatology, Paleoecology* **147**: 73-99.

Meehan, B. 1982). *Shell Bed to Shell Midden.* Australian Institute of Aboriginal Studies, Canberra.

Melim, L.A. 1996). Limitations on lowstand meteoric diagenesis in the Pliocene-Pleistocene of Florida and Great Bahama Bank: Implications for eustatic sea-level models. *Geology* **24(10)**: 893–896.

Morris, K.D. (1991). *Management Proposals for the Montebello Islands and Surrounding Waters. CALM Occasional Paper No. 3/91 August 1991.* Department of Conservation and Land Management, Como.

Morse, K. (1988). Mandu Mandu Creek Rockshelter: Pleistocene human occupation of North West Cape, Western Australia. *Archaeology in Oceania* **23**:82–88.

Morse, K. 1993a). New radiocarbon dates from North West Cape, Western Australia: a preliminary report. pp. 204–213. *In*, Smith, M.A., Spriggs, M. and Fankhauser, B. (eds), *Sahul in Review: The archaeology of Australia, New Guinea and Island Melanesia. Occasional Papers in Prehistory* Number **31**. Department of Archaeology and Natural History, Research School of Pacific Studies, The Australian National University Press, Canberra.

Morse, K. 1993b). *West Side Story: Towards a Prehistory of the Cape Range Peninsula, Western Australia.* Unpublished PhD Thesis, Centre for Prehistory, University of Western Australia, Perth.

Morse, K. 1996). Coastal shell middens, Cape Range Peninsula, Western Australia: an appraisal of the Holocene evidence. pp. 9–25. *In*, Veth, P. and Hiscock, P. (eds), *Archaeology of Northern Australia: regional perspectives. Tempus* Number **4**. University of Queensland Press, St Lucia.

Morse, K. (1999). Coastwatch: Pleistocene resource use on the Cape Range Peninsula. pp. 73–80. *In*, Hall, J. and McNiven, I.J. (eds), *Australian Coastal Archaeology.* ANH Publications, Research School of Pacific and

Asian Studies, The Australian National University, Canberra. *Research Papers in ANH* Number **31**.

Morton, S. and Baynes, A. (1998). Small mammal assemblages in arid Australia: A reappraisal. *Australian Mammalogy*, **8**:159-169.

Mowat, F. 1994). Size really does matter. pp. 201–209. *In*, Sullivan, M., Brockwell, S. and Webb, A. (eds), *Archaeology of the North: proceedings of the 1993 Australian Archaeological Association Conference.* Northern Australian Research Unit, The Australian National University, Darwin.

Mulvaney, J. and Kamminga. J. (1999). *Prehistory of Australia.* Allen and Unwin, St Leonards.

Mylroie, J.E. 2001). Karst features of Guam in terms of a general model of carbonate island karst. *Journal of Cave and Karst Studies* **63(1)**: 9–22.

O'Connor, S. (1992). The timing and nature of prehistoric island use in northern Australia. *Archaeology in Oceania* **27**: 49–60.

O'Connor, S. (1994). A 6700 bp date for island use in the West Kimberley, Western Australia: New evidence from High Cliffy Island. *Australian Archaeology* **39**: 102-108.

O'Connor, S. (1999a). *30,000 Years of Aboriginal Occupation: Kimberley, North West Australia. Terra Australis* Number **14**. ANH Publications, Research School of Pacific and Asian Studies, The Australian National University, Canberra.

O'Connor, S. (1999b). A diversity of coastal economies: Shell mounds in the Kimberley region in the Holocene. pp. 37–50. *In*, Hall, J. and McNiven, I.J. (eds), *Australian Coastal Archaeology. Research Papers in ANH* Number **31**. ANH Publications, Research School of Pacific and Asian Studies, The Australian National University, Canberra.

O'Connor, S. and Sullivan, M. 1994). Distinguishing middens and cheniers: a case study from the southern Kimberley, WA. pp. 26–49. *In*, Veth, P. and Hiscock, P. (eds), *Archaeology of Northern Australia: regional perspectives. Tempus* Number **4**. University of Queensland Press, St Lucia.

O'Connor, S. and Veth, P. 1993). Where the desert meets the sea: a preliminary report of the archaeology of the Kimberley coast. *Australian Archaeology* **37**: 25–34.

O'Connor, S. and Veth, P. (2000). The world's first mariners: Savannah dwellers in an island continent. pp. 99–137. *In*, O'Connor, S. and Veth, P. (eds), *East of Wallace's Line: Studies of Past and Present Maritime Cultures of the Indo-Pacific Region. Modern Quaternary Research in Southeast Asia* Number **16**. A.A. Balkema, Rotterdam.

O'Connor, S., Veth, P. and Barham, A.J. (1999). Cultural versus natural explanations for lacunae in Aboriginal occupation deposits in northern Australia. *Quaternary International* **59**: 61-70.

O'Connor, S., Veth, P. and Campbell, C. (1998). Serpent's Glen Rockshelter: Report of the first Pleistocene-aged occupation sequence form the Western Desert. *Australian Archaeology* **46**: 12–22.

Przywolnik, K. (2002a). Coastal sites and severe weather in Cape Range Peninsula, northwest Australia. *Archaeology in Oceania* **37**(3): 137–52.

Przywolnik, K. (2002b). *Patterns of Occupation in Cape Range Peninsula (WA) over the last 36,000 Years.* Unpublished PhD thesis, Department of Archaeology, University of Western Australia, Nedlands.

Przywolnik, K. (2005). Longterm transitions in hunter-gatherers of coastal northwestern Australia. pp. 177–205. *In*, Veth, P., Smith, M. and Hiscock, P. (eds), *Desert Peoples: Archaeological perspectives.* Blackwell Publishing, Oxford.

Pulsford , T. 1994). *Montebello Repast: Early Holocene environments and economy on the Montebello Islands—Pilbara desert coast, Western Australia.* Unpublished BA(Hons) Thesis, Department of Archaeology and Anthropology, James Cook University, Townsville.

Scall, A. (1985). *Aboriginal Use of Shell on Cape York. Cultural Resource Management Monograph* **6**. Department of Community Services, Brisbane.

Scurla, S. (1996). *The Stefano Castaways* (2nd ed.) [translated and edited by Sala, A.] Perth. [Original manuscript, *I Naufraghi dello Stefano,* 1876]

Sedgewick, E.H. (1978). A population study of the Barrow Island avifauna. *The Western Australian Naturalist* **14**(4): 85–108.

Semeniuk, V. 1993). The Pilbara Coast: a riverine coastal plain in a tropical arid setting, northwestern Australia. *Sedimentary Geology* **84**: 235–256.

Semeniuk, V. 1996). Coastal forms and Quaternary processes along the arid Pilbara coast

of northwestern Australia. *Paleogeography, Paleoclimatology, Paleoecology* **123**: 49–84.

Shackley, M. (1981). *Environmental Archaeology.* Allen and Unwin, London.

Short, J. and Turner, B. 1991). Distribution and abundance of spectacled hare-wallabies and euros on Barrow Island, Western Australia. *Wildlife Research* **18**: 421–429.

Short, J. and Turner, B. (1992). The distribution and abundance of the banded and rufous hare-wallabies, *Lagostrophus fasciatus and Lagorchestes hirsutus. Biological Conservation* **60**: 157–166.

Short, J. and Turner, B. (1993). The distribution and abundance of the burrowing bettong (Marsupialia : Macropodidae). *Wildlife Research* **20**: 525-534.

Short, J., Richards, J.D. and Turner, B. (1998). Ecology of the Western Barred Bandicoot (*Perameles bougainville*) Marsupialia: Peramelidae on Dorre and Bernier Island, Western Australia. *Wildlife Research* **25**: 567–586.

Sim, R. (1998). *The Archaeology of Isolation? Prehistoric occupation in the Furneaux Group of Islands, Bass Strait, Tasmania.* Unpublished PhD Thesis, Research School of Pacific and Asian Studies, The Australian National University, Canberra.

Sim, R. (2002). Preliminary Results from the Sir Edward Pellew Islands Archaeological Project, Gulf of Carpentaria 2000-2001. Unpublished report to the Australian Institute of Aboriginal and Torres Strait Islander Studies.

Sim, R. (2004). Holocene palaeoenvironmental change and Australian Aboriginal occupation patterns on offshore islands: A case study from Vanderlin Island in the Gulf of Carpentaria. Unpublished paper delivered at the Global Perspectives on the Archaeology of Islands, Auckland, New Zealand, December 2004.

Slack-Smith, S. (1993). The non-marine molluscs of the Cape Range Peninsula, Western Australia. pp. 87–108. *In*, Humphreys, W.F. (ed.), *The Biogeography of Cape Range, Western Australia. Records of the Western Australian Museum Supplement* **45**.

Smith, D.N. 1965). *Montebello Geological Reconnaissance.* Unpublished report to Western Australian Petroleum Pty Ltd.

Smith, M.A. and Sharp, N.D. (1993). Pleistocene sites in Australian, New Guinea and Island Melanesia: Geographic and temporal structure of the archaeological record. Pp. 37-59. *In*, Smith, M.A., Spriggs, M., and Fankhauser, B. (eds), *Sahul in Review: The Archaeology of Australia, New Guinea and Island Melanesia*. Department of Prehistory, Research School of Pacific and Asian Studies, The Australian National University, Canberra.

Smith, M.A. and Veth, P. (2004). Radiocarbon dates for baler shell in the Great Sandy Desert. *Australian Archaeology* **58**: 37-38.

Smith, P. (2000). Dietary stress or cultural practice: Fragmented bones at the Puntutjarpa and Serpent's Glen rockshelters. *Australian Archaeology* **51**: 65–66.

Start, A.N. and Kitchener, D.J. (1998). Western Pebble-mound Mouse. pp. 590-592. *In*, Strahan, R. (ed.), *The Mammals of Australia* (2nd ed.). Reed Books, Chatswood.

Storr, G.M., Smith, L.A. and Johnstone, R.E. (2002). *Snakes of Western Australia* (Revised ed.). Western Australian Museum, Perth.

Strahan, R. 1998). *The Mammals of Australia* (2nd ed.). Reed Books, Chatswood.

Tindale, N.B. 1974). *Tribal Boundaries in Aboriginal Australia*. Division of National Mapping, Canberra.

Tonkinson, R. 1978). *The Mardudjara Aboriginies living the dream in Australia's desert.* Holdt, Rinehart and Winston, New York.

Turnbridge, D. 1991). *The Story of the Flinders Ranges Mammals.* Kangaroo Press, Kenthurst.

van der Kaars, S.A. (1991). Palynology of eastern Indonesian marine piston-cores: A late Quaternary vegetational and climatic record for Australasia. *Paleogeography, Paleoclimatology, Paleoecology* **85**: 239–302.

van der Kaars, S. and De Deckker, P. (2002). A Late Quaternary pollen record from deep-sea core Fr10/95, GC17 offshore Cape Range Peninsula, northwestern Western Australia. *Review of Palaeobotany and Palynology* **120**: 17-39.

van der Kaars, S., Wang, X., Kershaw, P., Guichard, F. and Setiabudi, D.A. (2000). A late Quaternary palaeoecological record from the Banda Sea, Indonesia: patterns of vegetation, climate and biomass burning in Indonesia and northern Australia. *Paleogeography, Paleoclimatology, Paleoecology* **155**: 135–153.

Vardon, M.J. and Tidemann, C.R. 1997). Black flying foxes, *Pteropus alecto*: Are they different in North Australia? *Australian Mammology* **20**: 131–133.

Veitch, B. (1999). Shell middens on the Mitchell Plateau: A reflection of a wider phenomenon? pp. 51–64. *In*, Hall, J. and McNiven, I.J. (eds), *Australian Coastal Archaeology. Research Papers in ANH* Number **31**. ANH Publications, Research School of Pacific and Asian Studies, The Australian National University, Canberra.

Veth, P. 1993). The Aboriginal occupation of the Montebello Islands, north-west Australia. *Australian Aboriginal Studies* **24**: 81–89.

Veth, P. (1995). Aridity and settlement of northwest Australia. *Antiquity* **69**: 733–746.

Veth, P. (1999). The occupation of arid coastlines during the terminal Pleistocene of Australia. pp. 65–72. *In*, Hall, J. and McNiven, I.J. (eds), *Australian Coastal Archaeology. Research Papers in ANH* Number **31**. ANH Publications, Research School of Pacific and Asian Studies, The Australian National University, Canberra.

Veth, P. (2005). Cycles of aridity and human mobility: risk-minimization amongst late Pleistocene foragers of the Western Desert, Australia. pp.100-115. *In*, Veth, P., Smith, M. A. and P. Hiscock (eds), *Desert Peoples: Archaeological Perspectives*. Blackwell Publishing, Oxford.

Veth, P. and O'Brien, B. 1986). Middens on the Abydos Plain, northwest Australia. *Australian Archaeology* **22**: 45–59.

Veth, P., Bradshaw, E., Gara, T., Hall, N., Haydock, P. and Kendrick, P. (1993). *Burrup Peninsula Aboriginal Heritage Project*. Unpublished report to the Department of Conservation and Land Management, Como.

Veth, P., O'Connor, S. and Wallis, L.A. (2000). Perspectives on ecological approaches in Australian archaeology. *Australian Archaeology* **50**: 54–66.

Veth, P., Smith, M. and Hiscock, P. (eds) (2005). *Desert Peoples: Archaeological Perspectives*. Blackwell Publishers, Oxford.

Veth, P. and McDonald, J. (in prep). Dating of Bush Turkey Rock Shelter 3, Calvert Ranges, Establishes Early Holocene Occupation of the Little Sandy Desert, Western Australia.

Vinnicombe, P. (1987). *Dampier Archaeological Project: Resource documentation, survey and salvage of Aboriginal sites, Burrup Peninsula, Western Australia*. Unpublished report to the

Department of Aboriginal Sites, Western Australian Museum, Perth.

Watts, C.H.S. and Aslin, H.J. 1981). *The Rodents of Australia*. Angus and Robertson, Sydney.

Webster, P.J. and Streten, N.A. 1978). Late Quaternary ice-age climates of tropical Australasia: Interpretations and reconstructions. *Quaternary Research* **10**: 279–309

Wells, F.E. and Bryce, C.W. 1988). *Seashells of Western Australia*. Western Australian Museum, Perth.

Wells, F., Slack-Smith, S. and Bryce, C.W. (1994). Molluscs. *In*, Berry, P.F. (ed), *Survey of the Marine Fauna and Habitats of the Montebello Islands, August 1993*. Western Australian Museum, Perth.

Wicks, C.M. and Herman, J.S. 1995). The effect of zones of high porosity and permeability on the configuration of the saline-freshwater mixing zone. *Groundwater* **33(5)**: 733–740.

Wilson, B. (2002) *A Handbook of Australian Seashells on Seashores East to West and North to South*. Reed New Holland, Sydney.

Withers, P.C. and O'Shea, J.E. 1993). Morphology and physiology of the Squamata. pp. 172–196. *In*, Glasby, C.J., Ross, G.J.B. and Beesley, P.L. (eds), *Fauna of Australia Volume 2A Amphibia and Reptilia*. Australian Government Publishing Service, Canberra.

Woodroffe, C.D., Chappell, J., Thom, B.G. and Wallensky, E. (1989). Depositional model of a macrotidal estuary and floodplain, South Alligator River, Northern Australia. *Sedimentology* **36**: 737–756.

Wyrwoll, K.H., Kendrick, G.W. and Long, J.A. (1993). The geomorphology and late Cenozoic geomorphological evolution of Cape Range–Exmouth Gulf region. pp. 1–24. *In*, Humphreys, W.F. (ed.), *The Biogeography of Cape Range, Western Australia. Records of the Western Australian Museum Supplement* **45.**

Yokoyama, Y., Lambeck, K., De Deckker, P., Johnston, P. and Fifield, L.K. (2000). Timing of the Last Glacial Maximum from observed sea-level minima. *Nature* **406**: 713-716.

APPENDIX I

Notes on biogeographically significant mammalian taxa recovered from Noala and Hayne's Caves, Montebellos Islands

Ken P. Aplin

Several of the mammal species identified in Noala and Hayne's represent significant new additions to the late Quaternary fauna of northwest Australia. This appendix provides information on key specimens in support of the taxonomic determinations of two taxa, *Onychogalea unguifera* and *Bettongia lesueur*.

Onychogalea unguifera (Gould, 1841)

The remains of this species were found only in the Hayne's Cave deposit. The most complete specimen is a partial right dentary retaining the following erupted teeth: lower incisor, the deciduous P_{2-3}, M_{2-3} (Fig AI.1). The unerupted P_3 was exposed by careful excavation of the buccal side of the dentary. The small, deeply bifid P_3 is highly diagnostic for the genus *Onychogalea* and all other elements of the dentition are consistent with this determination. The cheekteeth of this specimen and several others that retain teeth (Table AI.1) are slightly smaller than those of modern specimens of *O. unguifera* from the east Kimberley region of Western Australia (specimens in Australian National Wildlife Collection, CSIRO, Canberra; prefix CM). They substantially exceed those of *O. lunata* and *O. fraenata*, the only other known species of this genus.

Bettongia spp.

Regionally, there are records of only one species of this genus—the burrowing bettong or boodie, *Bettongia lesueur* (Quoy & Gaimard, 1824). This species survives today on Barrow Island but is recorded as a sub-fossil on Northwest Cape. It occurred historically through much of the woodland and shrubland vegetation zones of southwestern Australia, north to the Shark Bay region where it survives on Bernier and Dorre Islands. Other populations are known to have inhabited more arid regions of Western and Central Australia and semiarid regions of southeastern Australia. A second bettong, the woylie (*B. penicillata* Gray, 1837) is found in southwestern Australia, extending across southern Australia (Wakefield, 1967). Subfossil specimens of *B. penicillata* are recorded north to Faure Island in the southern end of Shark Bay. A very small bettong species, *B. pusilla* McNamara, 1997, is known only from subfossil remains from the Nullarbor Plain.

All of the archaeological specimens that retain the premolar teeth or their alveoli show these teeth to be aligned with the long axis of the molar row (Fig. AI.2), as in *Bettongia lesueur*, *B. tropica* Wakefield, 1867 and *B. gaimardi* (Desmarest, 1822), rather than flexed outward anteriorly as in *B. penicillata*. Of the former group, only *B. lesueur* has a geographic range that includes northwest Australia (the other species are restricted to eastern Australia). It thus seems likely that all of the archaeological material is referrable to *B. lesueur sensu lato*. However, as noted in the main text, the pooled sample is more variable in tooth size and morphology than would be expected of a single species of this genus.

Selected cranial and cheektooth dimensions of various modern and historical populations of *B. lesueur* are summarized in Table AI.2. These illustrate the striking difference in cranial and cheektooth dimensions and proportions between the Barrow Island population and those located further south (Bernier Island, Dorre Island and mainland populations). This contrast is observed in many aspects of external and cranial morphology—the Barrow Island population is smaller bodied and shorter tailed, has a distinctive fur colouration, and the cranium has a shorter rostrum and a steeper molar gradient (i.e. posterior molars smaller relative to anterior molars). Compared with the Bernier and Dorre Island populations, the cranium of Barrow Island specimens also features proportionally larger auditory bullae. However, the bullae are larger again in mainland southwest populations which are sometimes distinguished at subspecific level [as *B. lesueur grayi* (Gould, 1841)] from those on the Shark Bay islands (*B. lesueur lesueur*). The molars of the Barrow Island population are also less bulbous basally than those of the southern populations, such that individual teeth can be distinguished provided they are not too heavily worn. Considering all of these factors, it seems probable that the Barrow Island population is genetically distinct from that found in more southerly parts of Western Australia, albeit clearly a member of a *B. lesueur* 'species group'. Surprisingly, no taxonomic name has been proposed for this distinctive population.

Measurements of the Montebellos archaeological *Bettongia* (Table AI.3) span the combined range of dental variation of both the Shark Bay and

Barrow Island populations. This is clearly illustrated in bivariate plots of measurements for individual teeth (Fig. AI.3). Although the bivariate plots suggest a continuum of size variation (i.e. the two clusters abut or overlap), as noted above the morphology of the molar teeth provides a further basis for assigning them to either the 'large' or 'small' form. The subfossil specimens also include characteristically 'gracile' and 'robust' skeletal elements such as the figured dentaries (Fig. AI.4).

Two interpretations might be advanced to account for the observed variation. First, it is possible that during the late Pleistocene and early Holocene the Barrow Island *B. lesueur* was more variable in tooth size than any of surviving or historical populations. Reduction in size variation and overall size diminution might be expected in the Barrow Island population following sea level rise. However, the same reasoning would not account for the comparably low level of variation observed in the historical mainland population in southwestern Australia (and in other *Bettongia* species that show similar levels of dental variability). Alternatively, it is possible that two distinct lineages of *B. lesueur*, the 'Barrow Island' and 'southwestern' forms of today, co-existed in the Montebello region during the late Pleistocene and early Holocene. Based on their current distributions and patterns of habitat use, it seems likely that they would occupy distinct habitats, the smaller Barrow Island bettong using the rocky limestone plateau and the southwestern bettong—or boodie—occupying the sand plain habitats of the exposed continental shelf (Short and Turner, 1993). With marine transgression, the latter taxon would have contracted in range to the southern part of the Western Australian coastline.

For this work, we have favoured the latter interpretation and discuss the 'Barrow Island' and 'southwestern' forms of *Bettongia lesueur* as though they represent distinct species. One additional reason for favouring this interpretation is that the relative abundances of the two forms differ between and within each of the two deposits—the 'Barrow Island' form is more abundant in the Noala Cave deposit and in the uppermost levels of Hayne's Cave, while the 'southwestern' form is more abundant in the rich midden layers of Hayne's Cave.

LITERATURE CITED

McNamara, J.A. (1997). Some smaller macropod fossils of South Australia. *Proceedings of the Linnean Society of New South Wales*, **177**: 97-106.

Wakefield, N.A. (1967). Some taxonomic revision in the Australian marsupial genus *Bettongia* (Macropodidae), with description of a new species. *The Victorian Naturalist*, **84**: 8-22.

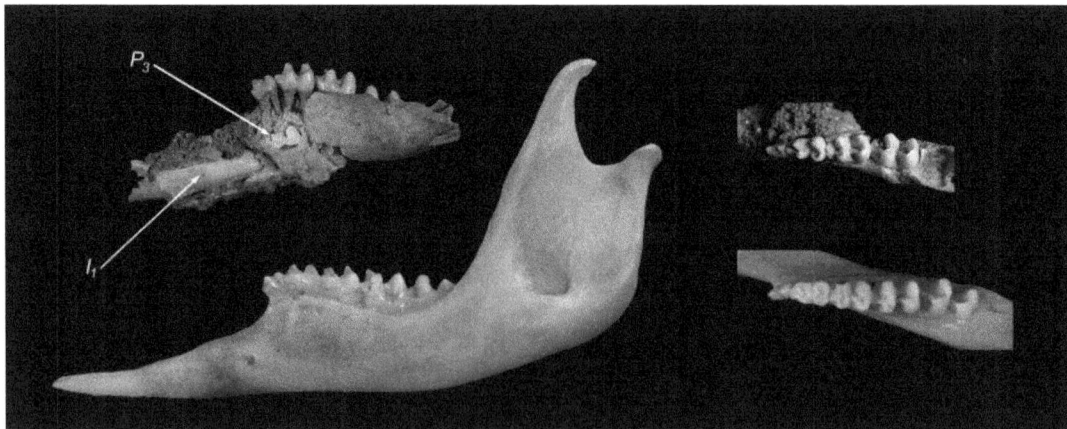

Figure AI.1 Sub-fossil fight dentary of Onychogalea unguifera from Hayne's Cave, Montebellos Islands (spit HC-4-9) compared with a modern specimen of this taxon from near Kununurra (CM15390). The sub-fossil specimen has the lower incisor, two deciduous premolars and two molars in occlusion. The unerupted permanent premolar (P₃) has been exposed by excavation of the overlying bone. The modern specimen is at a slightly later stage of dental eruption with the third permanent molar also in occlusion.

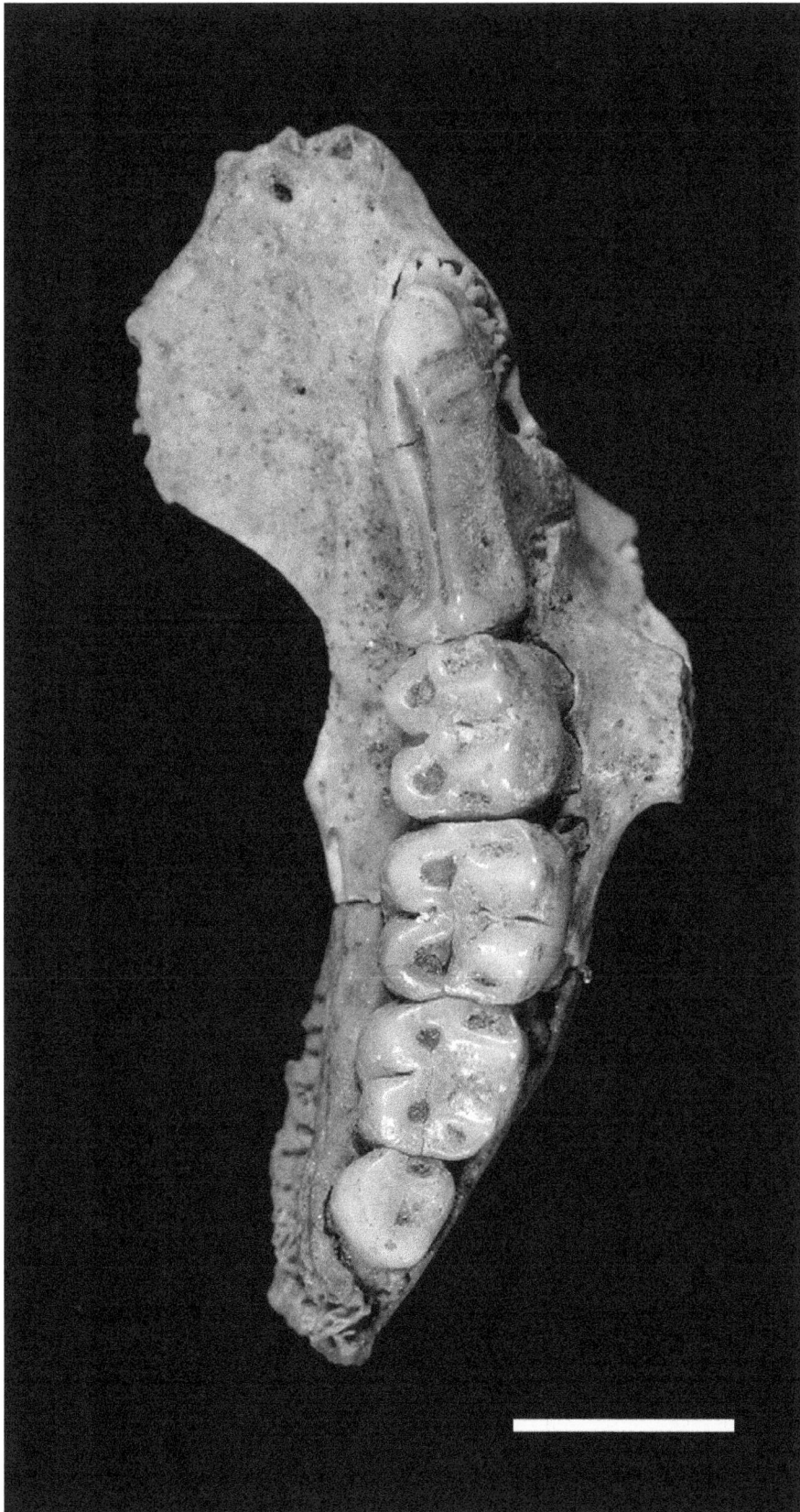

Figure AI.2 Subfossil left maxilla of Bettongia lesueur ('southwestern' form) from Hayne's Cave (spit HC-4-9) illustrating the lack of antero-buccal flexion of the permanent premolar (P_3) relative to the long axis of the molar row (M_{1-4}).

Veth, Aplin, Wallis, Manne, Pulsford, White and Chappell

UPPER M2

UPPER M3

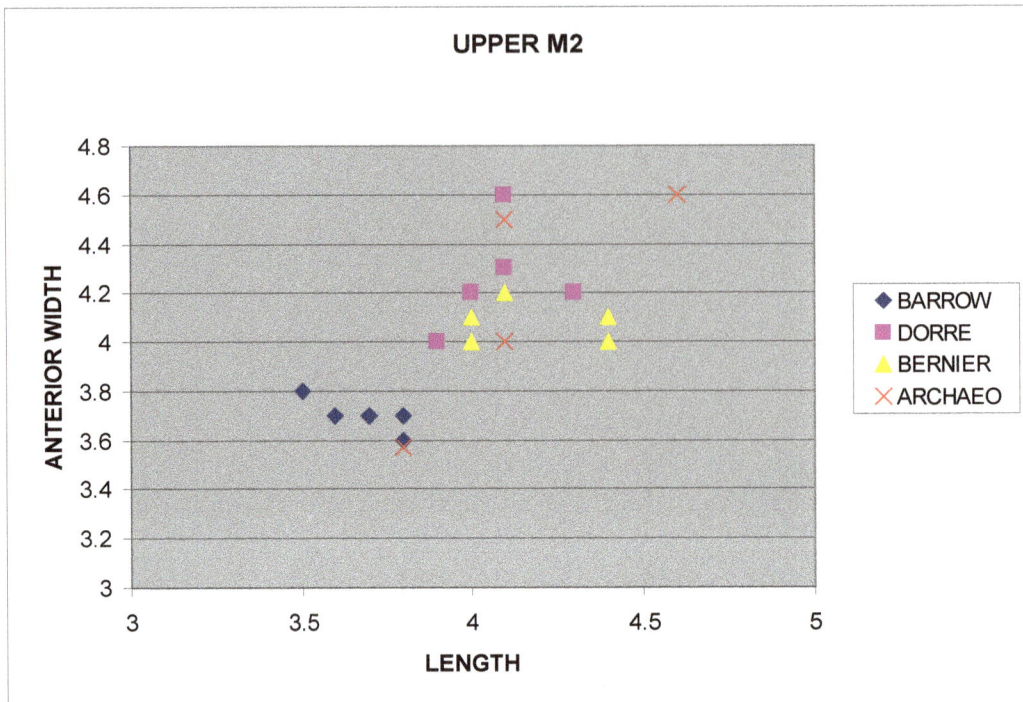

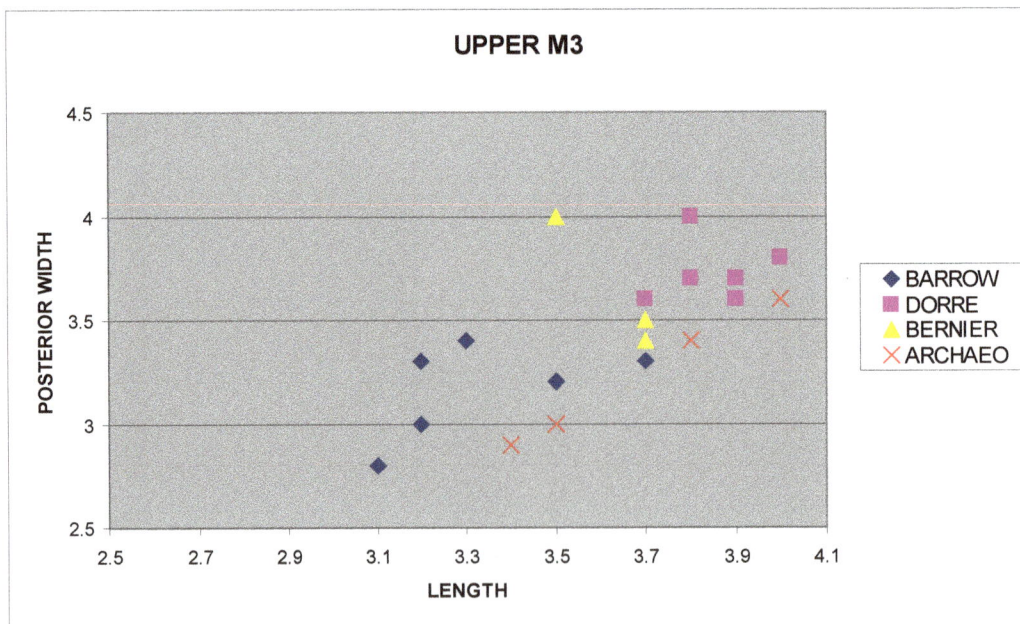

Figure AI.3 Bivariate plots illustrating the differences in molar size between the various Island populations of Bettongia lesueur and the position of several specimens from Hayne's and Noala Caves. Illustrated are the length vs anterior widths for each of M^2 and M^3 (see Table AI.3 for data). The Hayne's Cave sub-fossil sample spans almost the entire range of variation of the combined island populations.

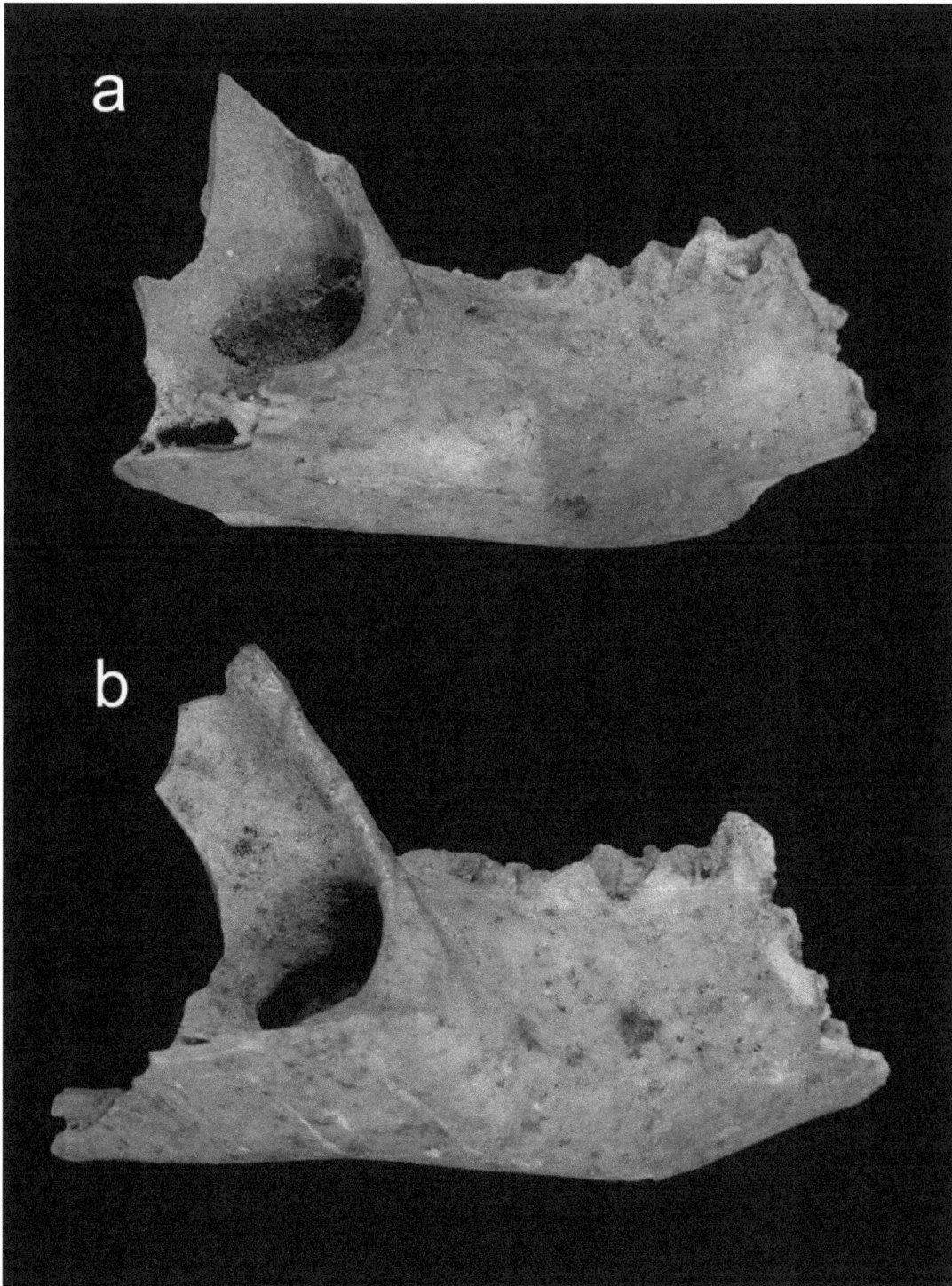

Figure AI.4 Subfossil edentulous right dentaries from Hayne's Cave (both from spit HC-4-10) representing a) the 'Barrow Island' and b) the 'southwestern' form of Bettongia lesueur. Both specimens are adult with all molars fully erupted. The latter is more robust, with a more prominent ventral flexion and larger opening for the masseteric canal. The two specimens were photographed at the same scale.

Veth, Aplin, Wallis, Manne, Pulsford, White and Chappell

Table A1.1 Measurements (in mm) of the lower cheekteeth of three sub-fossil specimens of Onychogalea unguifera from Hayne's Cave, Montebello Islands and one modern specimen (CM15390: from 8 km east of Kalumburu Mission, W.A.).

Spit	Symmetry	dp_2L	dp_2W	dp_3L	dp_3AW	dp_3PW	P_3L	P_3W	M_1L	M_1AW	M_1PW	M_2L	M_2AW	M_2PW	M_3L	M_3AW	M_3PW
HC-4-3	L								4.8	2.6	3.2	5.5	3.6	3.6			
HC-4-6	R			4.8	2.4	3.1	4.3	2.4									
HC-4-9	R	3.6	2.2	5.3	2.7	3.4						5.7	3.5	3.8	6.8	4.1	4.2
CM15390	L	3.7	2.1	5.0	2.9	3.2			5.2	3.5	3.9	6.3	4.6	4.9	7.2	5.3	

Table A1.2 Summary statistics for selected cranio-dental measurements (in mm) of various historical and contemporary populations of Bettongia lesueur, based on specimens in the mammal collection of the Western Australian Museum. Values in each cell are mean ± s.e., range and sample size. Only adult specimens with a fully erupted permanent cheektooth series (P^3-M^4) were included. Values for males and females are pooled due the presence of many unsexed specimens in the samples.

The measurements were taken as follows: Craniobasal Length (CBL): from occipital condyle to antero-ventral tip of premaxilla; Upper premolar length (P^3L): crown length of permanent upper premolar; Upper cheektooth row (P^3-M^4): crown length of permanent upper cheektooth row; Rostral depth (RD): vertical depth of rostrum measured midway along diastema; Nasal Width (NW): maximum combined width of nasal bones (at fronto-maxillary suture); Inter-bullar Width (IBW): minimum distance between auditory bullae.

	CBL	P^3L	P^3-M^4	RD	NL	IBW
Bernier ± Dorre Is	64.9 ± 0.373 57.6-67.3 34	8.8 ± 0.058 8.0-9.5 43	22.8 ± 0.12 21.0-24.9 44	12.8 ± 0.18 11.1-18.3 40	25.6 ± 0.29 21.9-28.6 26	9.8 ± 0.15 8.0-11.3 32
SW Mainland	68.6 ± 1.45 61.8-72.2 9	8.4 ± 0.18 7.4-9.3 11	23.5 ± 0.53 20.6-26.2 11	13.3 ± 0.34 10.4-11.5 10	27.7 ± 0.71 20.0-23.8 9	9.69 ± 1.20 7.9-11.3 9
Barrow Island	56.4 ± 0.437 54.0-57.8 8	7.6 ± 0.11 7.1-8.3 11	20.1 ± 0.20 18.9-20.9 11	11.0 ± 0.13 11.1-18.3 10	21.8 ± 1.14 20.0-23.8 9	9.1 ± 0.14 8.5-9.6 8

Table AI.3 Measurements (in mm) of the upper cheekteeth of selected individuals of three extant populations of *Bettongia lesueur* (Dorre, Bernier and Barrow Islands; WA Museum specimens, identified by registration number in the first column) compared with sub-fossil specimens from Hayne's and Noala Caves, Montebello Islands. The sex of modern specimens is shown where this information is known from an associated skin; pick-up skulls are indicated as unsexed (U). The symmetry (L/R) is shown for the sub-fossil specimens.

REGNO or SPIT	ISLAND	SEX	SYMMETRY	P3L	M1L	M2L	M3L	M4L	M1AW	M1PW	M2AW	M2PW	M3AW	M3PW	M4AW	M4PW
M17045	DORRE	M		9.1	4	4	3.7	2.3	4.1	4.3	4.2	4.2	3.8	3.3	2	2
M17052	DORRE	M		9.5	4	4.3	3.7	2.2	4	4	4.2	4	3.9	3.3	2.2	1.7
M17044	DORRE	M		9.1	4.1	4.1	3	2.9	4	4.4	4.6	4.5	4.1	3.5	2.9	2.3
M3645	DORRE	F		8.9	3.8	4	4	2.4	3.9	4	4.2	4	4.1	3.5	2.2	1.7
M3642	DORRE	F		8.7	3.9	3.9	3.5	2.4	3.7	3.6	4	3.7	3.8	3.2	2.7	1.7
M17042	DORRE	F		9.2	3.9	4.1	3.9	2.4	4.1	4.2	4.3	4.5	4.1	3.6	2.6	1.3
M17055	BERNIER	M		8.6	4.1	4.4	3.8	2.1	3.8	3.9	4	3.9	4	3.5	2.2	2.1
M5533	BERNIER	M														
M17056	BERNIER	F		8.2	3.7	4	3.4	2.3	4	3.9	4	3.6	3.5	3.1	2.5	2
M5470	BERNIER	F		8.8	3.9	4	3.8	2.2	4.2	4.4	4.4	3.9	3.7	3.1	2.1	1.3
M17057	BERNIER	F		8.5	3.9	4.4	3.7		4	3.7	4	3.8	3.8	3		
M5610	BERNIER	U		8.4	3.8	4.1	3.7	2.2	4.2	4.3	4.1	3.9	3.8	3.3	2.3	1.7
M7987	BERNIER	U		8.4	3.8	3.9	3.6	1.6	4	3.9	4.1	3.8	3.6	2.8	1.9	1.2
M5635	BERNIER	U		7.8	3.7	4.1	3.5	2	4.1	3.9	4.2	3.7	4.1	3.4	2.3	2
M7224	BARROW	M		7.7	3.6	3.8	3.3	1.5	3.3	3.3	3.6	3.6	3.2	2.6	1.6	
M7223	BARROW	F		7.1	3.2	3.5	3.2	1.7	3.6	3.8	3.8	3.4	3.3	2.7	2.4	2
M14875	BARROW	U		7.9	3.7	3.8	3.2	2.2	3.6	3.6	3.7	3.7	3.1	2.8	2.1	1.4
M14913	BARROW	U		7.7	3.5	3.5	2.9		3.5	3.6	3.8	3.7	3	2		
M13870	BARROW	U		8.1	3.5	3.7	2.8	1.4	3.6	3.6	3.7	3.7	2.8	2.3	1.4	1.4
M15488	BARROW	U		7.9	3.6	3.6	2.9	1.6	3.5	3.6	3.7	3.6	3	2.5		
HC-4-9	A		L	8.1	3.8	4.1	3.7	2.6	4.2	4.2	4.5	4.2	3.9	2.4	2.6	
HC-4-9	B		L	8.4												
HC-4-9	D		L	7.5	4.5	4.6	4.4		4	4.2	4.6	4.4	4.2	3.7		
HC-4-10	A		R			4.1					4	4				
HC-4-10	B		B	8.1												
HC-4-5	B		L	8												
HC-4-5	C		L	8.4												
NC-1-2			L			3.8					3.57	3.5				

Table AI.4 Measurements (in mm) of the lower cheekteeth of selected individuals of three extant populations of *Bettongia lesueur* (Dorre, Bernier and Barrow Islands; WA Museum specimens, identified by registration number in the first column) compared with sub-fossil specimens from Hayne's and Noala Caves, Montebellos Islands. The sex of modern specimens is shown where this information is known from an associated skin; pick-up skulls are indicated as unsexed (U). The symmetry (L/R) is shown for the sub-fossil specimens.

	ISLAND	SEX	SYMMETRY	P_3L	M_1L	M_2L	M_3L	M_4L	P_3AW	P_3PW	M_1AW	M_1PW	M_2AW	M_2PW	M_3AW	M_3PW	M_4AW	M_4PW
M17045	DORRE	M		7.4	3.8	4.2	4	2.4	3.6	2.8	3.3	3.7	3.9	4.2	4.1	3.8	2.6	1.7
M17052	DORRE	M		7.5	4	4.4	3.9	2.4	3.1	2.7	3.2	3.6	4	3.9	3.9	3.7	2.6	1.8
M17044	DORRE	M		7.1	4	4.1	3.9	2.9	3.7	2.7	3.4	3.7	3.8	3.9	3.9	3.6	3.2	2.4
M3645	DORRE	F		7	3.9	3.9	3.8	3.1	3.2	2.6	3.2	3.4	3.9	4.1	3.8	3.7	2.9	2.2
M3642	DORRE	F		6.7	3.3	4	3.7	2.6	2.8	2.5	2.9	3.4	3.6	3.7	4	3.6	2.8	2.3
M17042	DORRE	F		7.2	3.8	4.2	3.8	2.3	3.2	2.7	3.4	3.8	4.4	4.4		4	2.8	2.1
M17055	BERNIER	M		6.6	3.8	4.1	3.7	2.3	2.9	2.7	3	3.4	3.9	3.7	3.9	3.6	2.5	2.1
M5533	BERNIER	M		7	3.6	4.1	3.8	2.5	2.9	2.6	3.4	3.6	4.3	3.9	4	3.6	3	
M17056	BERNIER	F		6.8	3.4	4.1	3.7	2.9	3	2.6	3.2	3.7	3.8	3.7	3.6	3.4	2.8	2.3
M5470	BERNIER	F																
M17057	BERNIER	F		7	3.8	4	3.9	2.5	2.9	2.4	3.2	3.7	3.7	3.8	3.7	3.5	2.5	2.2
M5610	BERNIER	U		6.6	3.6	4	3.5	2.5	3	2.8	3.4	3.7	3.9	3.8	3.9	3.6	3.3	2.3
M7987	BERNIER	U																
M5635	BERNIER	U		6.8	3.6	4	3.8	2.2	3	2.8	3.2	3.6	4.1	4.2	4.3	4	2.9	1.4
M14875	BARROW	U		6.1	3.5	3.5	3.7	3.2	2.6	2.6	3	3.3	3.6	3.5	3.6	3.3	2.8	2.2
M7223	BARROW	F		5.7	3.6	3.8	3.3	2.6	2.7	2.2	3	3.3	3.8	3.8	3.9	3.4	3	2.2
M7224	BARROW	M		6.7	3.5	3.7	3.5	2.3	2.2	2	3.1	3.2	3.5	3.6	3.8	3.2	2.5	
M14913	BARROW	U		6.2	3.5	3.7	3.2	2.3	2.6	2.3	3	3.3	3.5	3.4	3.4	3	2.2	2.1
M13870	BARROW	U		6.6	3.5	3.6	3.1	2.4	2.3	2.3	3	3.3	3.6	3.4	3.2	2.8	2	1.6
M15488	BARROW	U		5.8	3.4	3.8	3.2	1.9	2.2	2.2	3	3.2	3.7	3.5	3.5	3.3	2.1	1.5
HC-4-9	C		R	7	4.3	4.6				2.6	3.6	4	4.3	4.2	4.1			
HC-4-5			R	7.5		3.9				3			3.8	3.4				
HC-4-3	A		R		4.1						3.4	3.6						
HC-4-3	B		L				3.5								3.5	3		
HC-4-5	A		L															
HC-4-5	D		R				3.8	2.8							3.8	3.4	2.5	1.9
HC-4-6	A		L															
HC-4-6	B		L		3.9		3.4				3.5	3.8			3.5	2.9		
NC-1-6	A		R															
NC-1-6	B		R			4.2							4	4				
NC-1-6	C		L			4.6							4.2	4				
NC-1-6	D		R				4								4.1	3.6		
NC-1-8					4.5						3.4	3.7						

62

Veth, Aptin, Wallis, Manne, Pulsford, White and Chappell

APPENDIX II:

Lithics from Noala Cave 1A and Haynes Cave 4, Montebello Islands

Elizabeth White

CONTENTS

1. INTRODUCTION TO DESCRIPTION OF LITHICS

This report provides a preliminary description and analysis of lithics from Noala Cave 1A and Haynes Cave 4.

Only a relatively small number of artefacts were recovered from the two excavations however they are potentially significant as they span a period of environmental change when dietary suites indicate a shift from terrestrial to marine resources. Such opportunities to investigate technological aspects of lithic assemblages correlating to changes in a resource base are not all that common.

The dating and character of the assemblages indicate that they fall within the expected range of generalised flake and core industries from north-western Australia which pre-date the introduction of hafted points and blades and other implements associated with technological shifts in the mid- to late Holocene (such as the introduction of the tula adze). General texts on such assemblages indicate a concern with typology, artefact size, raw material use and the identification of regionally distinct markers (Holdaway 1995, Lourandos 1997, Mulvaney and Kamminga 1999) with more detailed technical analyses and interpretations also considered (e.g. Hiscock 1988, 2002, McNiven 1994). The Montebello assemblages consist of a small number of artefacts (n=158) with only five retouched or modified artefacts. An analysis based on artefact type would not be particularly illuminating. Following Hiscock (1988:11ff) a technical approach has been adopted, considering the nature of the lithic material and attributes such as size, flake shape, platform angles and flake terminations.

A total of 19 artefacts from Noala Cave 1A, and 139 artefacts from Haynes Cave were identified from the lithic material recovered. It appears that natural pieces of tabular calcarenite were selected for the majority of the cores. These pieces may have initially been up to 10cm or more in size, with weights varying up to several hundred grams. The selected pieces may not have been particularly thick, however, with three of the four cores measuring 3cm or less in their third dimension. A larger single-platform core was thicker, measuring nearly 5cm. These stone pieces were reduced predominantly by unifacial (unidirectional) flaking, removing flakes from the shortest dimensions of the cores. Flaking was predominantly across the bedding plane of the stone, with core scars and most flaked debitage revealing horizontal banding. The larger single-platform core is twice as heavy as the other cores, with flakes struck from this core varying greatly in shape, with some flakes larger than those struck from the other cores.

Some change over time is suggested in the lithic assemblages. There are 15 artefacts from the lower unit of Noala Cave 1A associated with predominantly mammalian fauna while only 4 artefacts are associated with the upper assemblage associated with mammalian and marine fauna. The lower unit from Haynes Cave 4 associated with mammalian and marine fauna contained 62 artefacts while the assemblage associated with predominantly marine fauna had 77 artefacts.

There are some initial patterns evident from the assemblages linked to assemblages of terrestrial, mixed terrestrial and marine and then predominantly marine fauna from the sites which merit some comment on here. The smallest cores come from the lowest level of Noala Cave. Three of the 15 artefacts in the assemblage associated with terrestrial faunas have been made on exotic lithologies – two of quartz and one an ironstone. This indicates some kind of access to, trade of, stone resources from what is now the mainland. At the terminal Pleistocene groups may have been more residentially mobile. The technology associated with units demonstrating the utilisation of both mammal and marine resources includes larger cores and a wider variation in flake shape, possibly linked to a wider range of tasks resulting from the acquisition and processing of a broader dietary suite. If an expanded and richer resource base allowed people to be less mobile from time to time, then there may have been less constraint on tool kit portability, allowing the use of heavier cores. The presence of only a single fragment of a quartz retouched artefact (comprising <2% of the assemblage) suggests more restricted contact or access to stone resources on the mainland. The technology associated with the marine faunal assemblages was accompanied by smaller cores, yet larger flakes, and a more limited range of flake shapes tending to be less elongate. Relatively long narrow flakes found in the earlier unit may not have been optimal for tasks related to the processing of marine faunas (assuming they had some functional correlation). The smaller cores could relate to increased residential mobility patterns. There are no exotic lithics associated with the predominantly marine faunal assemblages.

2. ARTEFACT IDENTIFICATION

The lithics include items which can be positively identified as flaked artefacts and some items which appear to be broken pieces of stone. The identification of lithic items as flaked Aboriginal artefacts was recently considered by Holdaway and Stern at some length (2004:29ff). They suggested that identification should focus on (1) the identification of features characteristic of deliberate conchoidal fracture or bipolar flaking and (2) the context of the stone. Technical criteria based on stone fracture mechanics have

Veth, Aplin, Wallis, Manne, Pulsford, White and Chappell

been described by Speth (1972), Cotterell and Kamminga (1987) and Holdaway and Stern (2004), and discussed by Wright (1983).

A flake is a piece of stone which was struck from a larger rock. Flakes show specific technical features (Figure 1). Essentially, a flake has a platform, a point of impact (force application), a Hertzian cone, and a bulb of percussion. Flakes may also have lines resulting from shear fracture, a bulbar scar (also called *eraillure* scar) and ripple marks (Speth 1972:35). These features may have been more or less pronounced, depending on the quality of the stone material, the hardness of the hammer relative to the stone, and whether an anvil was used and the manner of its use.

Flakes were struck from larger rocks, usually identified as cores, tools or retouched artefacts. These artefacts show negative scars left after the detachment of flakes.

Flakes, cores and other modified items may have been broken, either during flaking or afterwards by trampling, sweeping or burning, or by natural weathering processes. If the side or end of a flake is broken the artefact is classified as a broken flake. Flakes broken longitudinally through their centres are classified as cone-split broken flakes. Sometimes only a fragment or piece of a flake was found (eg. a distal or medial piece) and these are classified as flake fragments. Some pieces of stone show signs of flaking but cannot be oriented towards a point of force application and these are classified as flaked pieces. Some artefacts are so broken that only remnant flaked surfaces remain and these are classified as remnant flaked artefacts.

It must be noted that the above discussions and description of technical features provide models for the technical identification of the products of flaking. In reality, flakes vary considerably in their shapes, and technical features are sometimes distorted by irregularities or flaws in the stone, or obscured by breakage or weathering. In practice it is not always possible to distinguish artefacts from other broken stone. The Montebello lithics include some items with fairly smooth surfaces which resemble those of flaked artefacts but lack other flaking traits. Such items have been noted as "possible artefacts" but are not included in the artefact totals.

The lithic items identified as artefacts from Noala Cave 1A and Haynes Cave 4 are sorted into categories based on the identification of fracture traits and breakage, as follows: core, other retouched fragment, flake, broken flake, flake fragment, flaked piece, remnant flake, possible artefact and other stone.

3. METHODS

The lithics from each site were sorted into spits and the following data recorded for each artefact. The data is entered into Microsoft's Access database programme, with the following fields underlined.

The site, square, spit and notes on separate bags are recorded as relevant.

Record. Each record in the database is assigned a number, to assist with management of the database and to enable reference to individual artefacts in the analysis.

Artefacts. The number of artefacts in each record.

PossibleArtefacts. The number of possible artefacts in each record.

NotArtefacts. The number of other lithic items in each record, usually the total number for each spit.

Lithology. The type of rock from which the artefact is made. Those of calcarenite are left blank, except for two from Haynes Cave which are of a fine grained stone, possibly similar to those from Noala Cave. Other types include quartz, FGS (other unidentified fine-grained siliceous stone) and other unidentified stone.

ArtefactType. Artefacts are sorted into types based on the identification of fracture traits and breakage, as described in section 2 above. The following categories are identified: core, retouched artefact, flake, broken flake, flake fragment, flaked piece, remnant flake, possible artefact and other stone.

Usewear? This field notes the possibility of wear from tool use.

MaximumSize. The maximum dimension of artefacts, in any direction. All artefacts were measured with calipers to the nearest 0.5mm.

WeightArtefacts. Weight is recorded for each artefact, to the nearest 0.1g for artefacts less than 50g in weight, to the nearest 0.2g for artefacts 50g-100g, and to the nearest 1.0 gram for artefacts 100g to 1kg.

WeightOther. Weight is recorded for all possible artefacts or other stone not identified as artefacts. The entered weight is for the items listed in the record; where more than one item is counted as a possible artefact or non artefact, the entered weight is the combined weight of all items.

Length. This is the oriented length of artefacts, measured from the point of force application and perpendicular to the platform for complete flakes, or maximum oriented length for broken flakes and flake fragments. Length is measured with calipers to the nearest 0.5mm. Broken dimensions are indicated by brackets. Negative scars on cores are measured in the same manner.

Width. This is the maximum width, measured perpendicular to length, and measured with calipers to the nearest 0.5mm. Broken dimensions are indicated by brackets. Negative scars on cores are measured in the same manner.

Thickness. This is the maximum artefact thickness, measured with calipers to the nearest 0.5mm.

Cores. Additional information is noted for cores to show how they were flaked, as advocated by Baker (1992). "Flaking pattern" is the pattern of flake removals evident on cores. The possible categories are unifacial, bifacial, asymmetric (including faceting) and bipolar (see Figure 2).

Platforms on flakes and broken flakes. The surface of platforms on flakes and broken flakes is noted. Potentially, seven types could have been present, although no faceted platforms are present (Figure 3).

- ❖ Cortex. Platform surface covered entirely with cortex.

- ❖ Plain. Platform surface consisting of a smooth flaked surface.

- ❖ Ridged. Platform surface has a ridge formed by a remnant margin of a flake formerly struck across the core.

- ❖ Scarred. Platform has one or a few flake scars, the points of force showing that they were initiated from blows struck on the dorsal edge of the platform surface.

- ❖ Faceted. Platform has many tiny flake scars on it, also initiated from the dorsal edge of the platform.

- ❖ Focal. A very small platform, less than twice the area of the ring crack.

- ❖ Indeterminate. The platform surfaces of some artefacts could not be determined due to irregularities on the platform surface, or because the platforms were partly collapsed or otherwise damaged.

Angle. Dorsal platform angle, measured on flakes and broken flakes, where possible at the flake's point of force application.

Termination. The distal end of flakes, where the ventral surface detaches from the core. Four types of terminations are identified - feather, hinge, step and plunging (the latter removing the base of the core). Step terminations may be difficult to distinguish from breaks.

Adhesions. Several of the artefacts have crystal growths on their surfaces, potentially affecting artefact weight and in some cases the identification of artefact type. A note is made in this field.

Comments. Any other comments about the artefacts, including cortex if present.

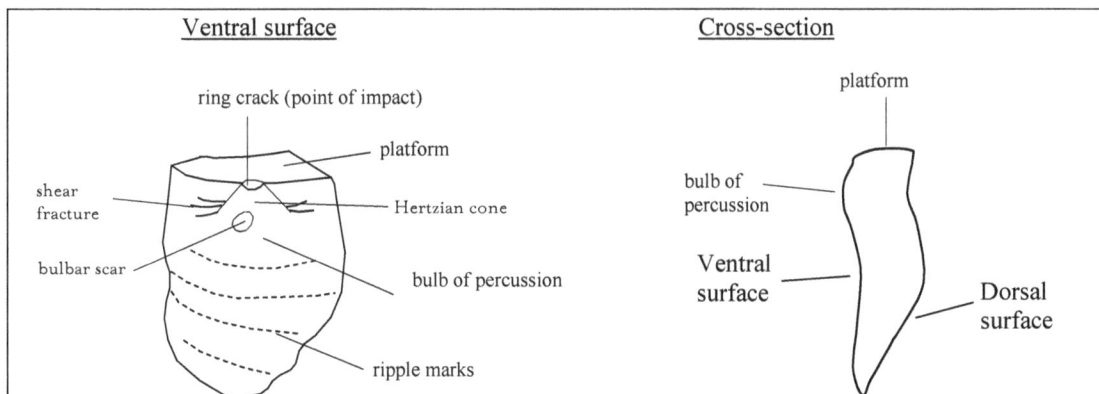

Figure 1: General features of a flake (after Speth 1972)

Unifacial flaking: Reduction proceeded from one edge on a face of a core (or tool). Cores may have been rotated, showing reduction from multiple faces but the force was applied in only one direction. On flakes, platforms may show cortical, plain smooth, or ridged platforms (see below).

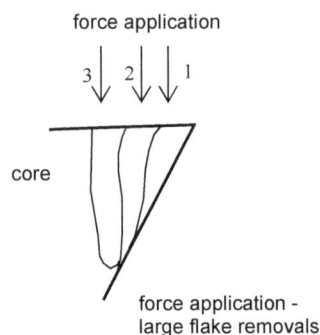

force application

3 2   1

core

force application -
large flake removals

Bifacial flaking: Relatively large-sized flakes (potentially useable as tools) were struck from the two faces of a platform edge. A bifacial pattern of removals made use of the bulbar scar from one flake removal to give a lower platform angle for a flake removed from the alternate face of the platform edge (Witter 1992:31); hence reduction proceeded from the two faces of a platform edge.

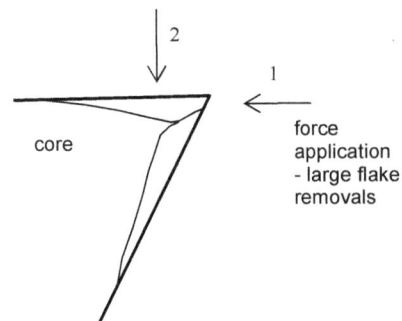

2

1

core

force
application
- large flake
removals

Asymmetric flaking: Small flakes (in the form of core preparation and platform faceting) were removed from the platform surface (1). Then (potentially useable) larger flakes (2) were struck from that prepared surface. Scarred or faceted platforms may result. Baker (e.g. 1992) argued that this flaking pattern is associated with blade production.

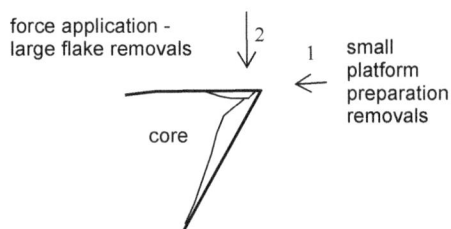

force application -
large flake removals

2

1   small
platform
preparation
removals

core

Bipolar: A reduction technique whereby the core is rested on an anvil and force applied to it an angle close to 90°, towards the core's contact with the anvil. Force passes through the core and bounces back from the anvil. The resulting flakes and core show crushing at the end which is struck by the hammer stone and at the end which was in contact with the anvil. The bipolar flakes may also have sheared or compressed bulbs of percussion (Cotterell and Kamminga 1987:688, 698-700).

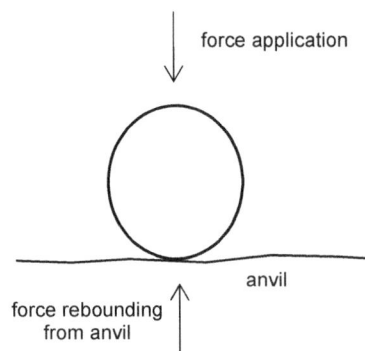

force application

anvil

force rebounding
from anvil

Figure 2: Core flaking patterns

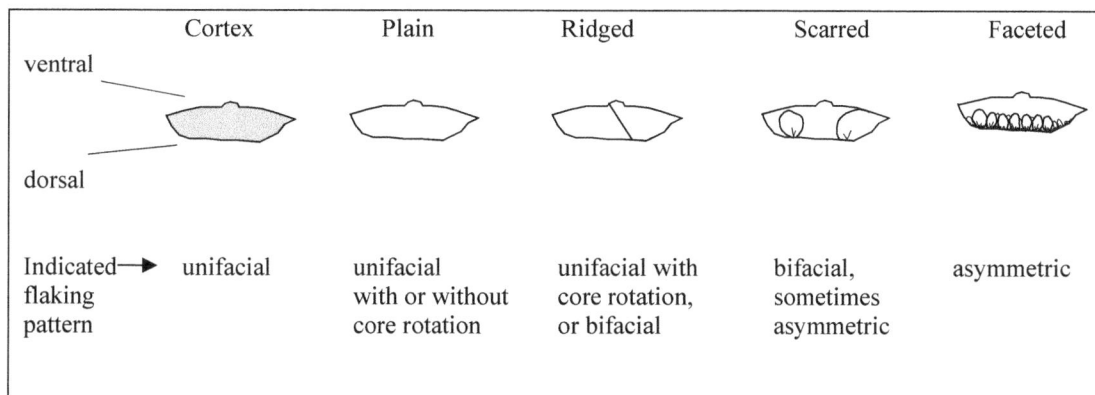

	Cortex	Plain	Ridged	Scarred	Faceted
ventral					
dorsal					
Indicated→ flaking pattern	unifacial	unifacial with or without core rotation	unifacial with core rotation, or bifacial	bifacial, sometimes asymmetric	asymmetric

Figure 3: Types of flake platform surfaces

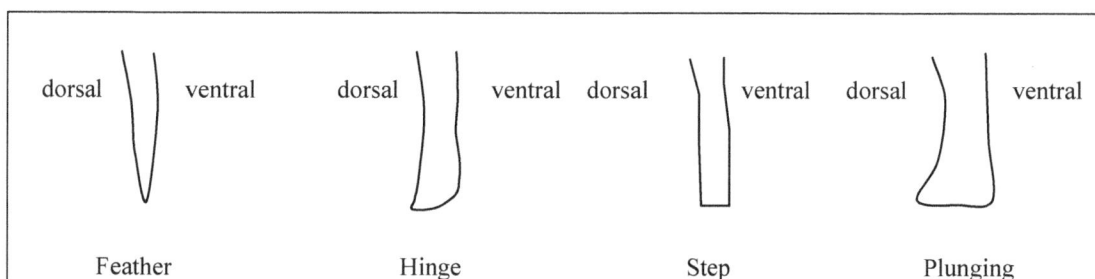

dorsal	ventral	dorsal	ventral	dorsal	ventral	dorsal	ventral
Feather		Hinge		Step		Plunging	

Figure 4: Types of terminations (cross section views, after Koettig 1994)

4. NOALA CAVE SQUARE 1A

A total of 48 lithics are present in the collection from Noala Cave square 1A. Of these, 19 can be identified as artefacts, 8 as possible artefacts and 21 others appear to be other pieces of (non cultural) broken stone. While artefact numbers are obviously low, two-thirds were recovered from spits 7 to 11, just above the uncalibrated age determination of 12,440 ± 110 BP (Wk-3432). Only four artefacts were recovered from spits 1 to 6, comprising largely dietary marine shell. "Other stone" not identified as artefacts is more frequent in the upper deposits.

Most of the artefacts appear to be of a fine-grained pink-cream-brown stone identified as calcarenite by MARP, with fine banding evident on the largest flake from spit 7 (Record 183) and on a few artefacts which were rinsed to check their identifications. This stone seem finer grained with sharper margins than much of the calcarenite (naturally) present in the Haynes Cave assemblage. It appears to have derived from a different source than the non-cultural lithic calcarenite recovered from Haynes Cave.

Three artefacts are of different stone materials. Two of milky and clear quartz are from spit 10. One is a flaked piece (20mm long) from a larger artefact, with a modified edge. The other is a fairly small flake fragment (14mm long). Another artefact of unusual stone is from spit 7. This is a flake of medium-grained ironstone, with cortex on most of its dorsal surface, indicating a pebble or cobble origin. Stratigraphically, these exotic lithologies relate to the "pre-midden" phase of occupation. It is likely that they were obtained from mainland sources – where volcanic bedrock is common.

Two cores are present – in spits 3 and 9. The core from spit 3 has a coating of sediment which obscures some technical details, but the artefact was not washed as it has some rounding to the platform edge that may be use-wear and have potential for residues. A broken piece of stone with remnant flaked scars from spit 5 may have been from another core. A conjoining flake indicates that the core from spit 9 was flaked on-site. The remaining artefacts are all *debitage* (Table 3).

Table 1: Lithics from Noala Cave square 1A

Marine shell	Spit	Identified artefacts	Possible artefacts	Other Stone
Frequent	1		1	
Numerous	2	1		3
	3	1		3
	4			4
	5	2		3
Present	6		1	4
Very rare	7	4		1
	8			
	9	3		
	10	2	3	1
	11	4		2
None	12	1	2	
	13	1	1	
	Total	19	8	21

Table 2: Noala Cave square 1A – artefacts of other lithologies

Lithology	Spit	No. of Artefacts
Ironstone?	7	1
Quartz	10	2

Table 3: Noala Cave 1A – artefact types

Spit	Cores	Flakes	Broken flakes	Flake fragments	Flaked pieces	Remnant flaked	Total artefacts
1							0
2						1	1
3	1						1
4							0
5						2	2
6							0
7		2		1		1	4
9	1			2			3
10				1	1		2
11		1	1	1		1	4
12			1				1
13		1					1
Total	2	4	2	5	1	5	19

5. HAYNES CAVE 4

A total of 139 artefacts were recovered from squares A and B, with another 12 possible artefacts and 181 other pieces of stone (Table 4). Artefact numbers are higher in square B (n=96) than in square A (n=43) indicating horizontal variation in the rates of lithic discard. The vertical distributions of artefacts are uneven – in square A modes occur in spits 2 and 3, spits 7 and 8, and spit 11; and in square B modes occur in spit 1, spits 3 and 4, and spits 6 to 8.

Most of the artefacts appear to comprise a fine to medium-grained cream to pink banded stone, sometimes weakly silicified and identified by MARP as calcarenite. Most of this stone differs from that in Noala Cave, appearing less siliceous with artefacts sometimes breaking along bedding planes.

Five of the artefacts which appear somewhat different may be due to variations within the general range of calcarenite quarries originally exploited (Table 5). There are accounts of 'darker' calcarenite outcrops occurring on Barrow Island (Ken Aplin, pers. obs.). Two broken flakes – one from square A spit 1 and the other from square B spit 7 – appear fine grained, and more like the lithology of those found in Noala Cave. An artefact from square A spit 12 is of grey-brown and orange-red stone, the grey-brown material possibly an inclusion within the orange-red stone. This material may be part of the range of variation within the calcarenite. Two small fragments from spits 6 and 9 are of glossy dark brown-grey stone and may be related to the grey-brown component of the artefact from spit 12.

Table 4: Lithics from Haynes Cave square 4A and 4B

Mammal fauna	Marine fauna	Spit	Square	Identified Artefacts	Possible Artefacts	Other stone	Square	Identified Artefacts	Possible Artefacts	Other stone
Present	Numerous	1	A	3			B	19		43
		2	A	7			B	5	1	29
Frequent	Abundant	3	A	6			B	18	5	21
		4	A	4			B	15	4	22
Numerous	Abundant	5	A	1	1		B	2		7 '
		6	A	2			B	12		13
	Abundant	7	A	4			B	13		13
		8	A	4			B	12	1	33
		9	A	3						
		10	A	2						
		11	A	4						
	Numerous	12	A	1						
	Present	13	A	2						
		Total		43	1			96	11	181

More exotic, however, is a fragment of a high quality retouched quartz artefact, found in square B spit 8. Its small size (11mm long and weighing 0.3g) indicates the discard of a remnant of little further utility, in keeping with the conservation of a rare type of stone (after Byrne 1980).

The assemblage from squares 4A and 4B include two cores, the quartz retouched fragment and *debitage* (Tables 5 and 6). The core from square A spit 9 is a large single platform core (a so-called 'horse-hoof' core).

Some variation between the assemblages from 4A and 4B is evident. Firstly, no cores were recovered from 4B, despite the presence of a larger number of artefacts. Secondly, artefact breakage rates are higher in 4B, with flake fragments making up 51% of the assemblage compared to 30% in 4A. Conversely, complete flakes make up only 12.5% of the assemblage in 4B compared to 23% in 4A.

Table 5: Haynes Cave square 4 – artefacts of other lithologies (FGS = fine grained siliceous)

Square	Lithology	Spit	No. of Artefacts
A	Like Noala Cave?	1	1
A	Dark FGS	6	1
A	Dark FGS	9	1
A	Dark & red FGS	12	1
B	Like Noala Cave?	7	1
B	Quartz	8	1

Table 6: Haynes Cave square 4A – artefact types

Spit	Cores	Flakes	Broken flakes	Cone-splits	Flake fragments	Total Artefacts
1			3			3
2		2	3		2	7
3	1	1	1		3	6
4		1	3			4
5					1	1
6					2	2
7		2		1	1	4
8		1	2		1	4
9	1			1	1	3
10		1	1			2
11		1	2		1	4
12					1	1
13		1	1			2
Total	2	10	16	2	13	43

Spit	Retouched fragment	Flakes	Broken flakes	Cone-splits	Flake fragments	Flaked pieces	Remnant flaked	Total Artefacts
1		1	1	2	8		7	19
2				1	2	1	1	5
3		1	3	1	11		2	18
4		1 + 1?	4	1	7	1		15
5		1	1					2
6		4	1	1	6			12
7		1	3	1	7		1	13
8	1	2			8	1		12
Total	1	12	13	7	49	3	11	96

6. REDUCTION TECHNOLOGY

This section is concerned with technical aspects relating to the reduction of lithic materials, as advocated by Hiscock (1988) and Holdaway (1995:792). To further investigate the technology in relation to changes in faunal evidence and chronology the lithics are analysed in four groups: (1) Haynes Cave spits 1 to 4, with predominantly marine faunal remains, (2) Haynes Cave spits 5 to 13, with both marine and mammalian faunal remains, (3) Noala spits 1 to 6, being earlier with both marine and mammalian faunal remains, and (4) Noala spits 7 to 13, the earliest of all, with predominantly mammalian faunal remains.

6.1 Cores

Four cores were recovered – two each from Noala Cave 1A and from Haynes Cave 4A. The core from Noala Cave 1A spit 3 has a coating of sediment which obscures some technical details. A flattish piece of stone was utilised measuring nearly 9cm across, 29mm in thickness, and weighing 213 grams. A flat surface was used as a platform and flakes removed unidirectionally from this, down the short dimension of the stone. Platform angles vary between 58° and 84°.

The core from Noala Cave 1A spit 9 is the smallest of the four, measuring under 6cm across and weighing only 52 grams (Figure 5). A triangular-shaped tabular piece of natural stone, 18mm thick, appears to have been utilised. Flakes were removed from the two flat surfaces, making use of the short dimension of the rock. The core was rotated and another flake was struck down the long dimension of the core, parallel to the bedding planes; this flake was recovered from spit 9 and conjoins to the core. Platform angles range between 60° and 87°.

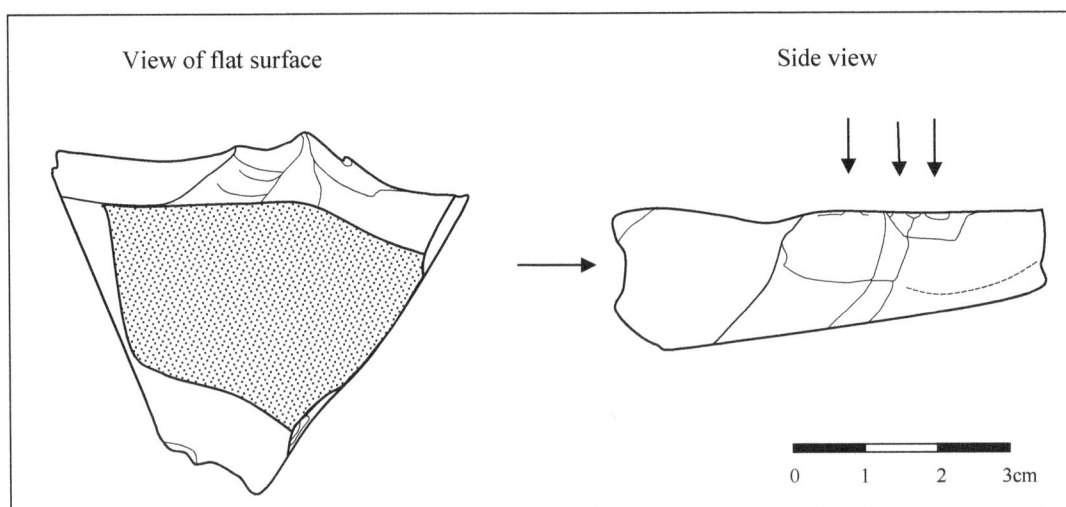

View of flat surface   Side view

Figure 5: Core from Noala Cave square 1A spit 9 (Speckling denotes natural surface, arrows indicate direction of removals.)

The core from Haynes Cave spit 3 has also been fashioned on a tabular piece of natural rock which is 22mm thick. It was originally bifacially flaked, resulting in a low angled platform (about 45°) which appears to have possible use-wear. The artefact broke, truncating the bifacial platform. The artefact was then reduced unifacially resulting in a higher platform angle (about 68°), removing flakes from the short axis of the core – perhaps an opportunistic change of technique in the reduction sequence. It now measures circa 11cm across, but was likely larger before breakage.

The core from square A spit 9 is a single platform ('horse-hoof') type, measuring 9cm across and is about 46mm thick. It was flaked unifacially from a flattish surface, with a few flakes removed bifacially from the surface of the main platform. Platform angles vary between about 80° and 90°. Some rounding of the platform edge could be due to weathering or usewear.

As a group the four cores share several attributes in common:

(1) They have been made from tabular pieces of stone, which may have formed naturally as a result of weathering of parent stone along bedding planes;

(2) The tabular pieces are not very thick, with three measuring between 18mm and 29mm in thickness; the single platform core is thicker at 46mm; and

(3) Flakes were removed predominantly by unidirectional flaking from a flattish platform down the short axis of the cores (that is, the shortest dimension of the core). Flaking was predominantly across the bedding plane of the stone, with flake scars tending to reveal the horizontal bands of the stone's (pre-metamorphosed) bedding planes.

The core from Noala Cave 1A spit 9 does have evidence for removal of flakes from the longer axis. This is also the smallest core recovered. A smaller core size may reflect higher mobility associated with hunting and foraging for terrestrial resources (after Kuhn 1994, Shott 1986).

The core from Haynes Cave 4A spit 3 also varied, having initially a low angle bifacial platform; the change to unifacial flaking may have occurred after the core broke.

Generally the four cores decrease in size (maximum dimension) over time, the smallest core being recovered from the deeper deposit at Noala Cave 1A and being associated with predominantly mammalian fauna. This may reflect decreasing residential mobility as marine habitats become proximal to the site(s). Obviously larger assemblages would be needed to test this proposition.

6.2 Artefact size

Overall, the majority of artefacts are less than 5cm in size, with only a few larger items (Table 8). The Haynes Cave assemblages are dominated by *debitage* less than 30mm in maximum dimension, with the two cores from 4A larger than this. The two cores from Noala Cave 1A are slightly smaller than those from Haynes Cave, measuring 60mm and 89mm[1]. The six artefacts more than 5cm in size include the four cores, a broken fragment with remnant flaked surfaces which may have derived from another core and a large flake.

The assemblage data was reanalysed, taking change in the faunal assemblages over time into account (Table 9). The assemblages from Haynes Cave associated with higher proportions of marine fauna have a larger number of artefacts. At Haynes Cave there was also an increase in the numbers of artefacts in the 3cm to 5cm size range: 25% of the assemblage from the deposits dominated by marine fauna compared to 10% of the previous assemblage contemporary with both mammal and marine fauna (excluding cores).

A chi-square test of significance on this variation in artefact size is significant at the 0.05 level for a two-tailed test ($X^2 = 5.215$, $df = 1$).

6.3 Artefact size and core scar size

The larger flake from Noala 1A indicates that some of the artefacts may have been struck from cores other than those recovered from the excavations. To address the issue the dimensions of scars on the cores were recorded and compared to the dimensions of flakes and broken flakes. The measurements were made with callipers to the nearest 0.5mm, in the same way that the artefacts were recorded. While many of the core scars are incomplete the shapes and locations of scars has made it possible to obtain an indication of the sizes and shapes of detached flakes.

---

[1] The other artefact from Noala 1A which is between 6cm and 7cm is a broken piece of stone with a remnant flaked surface, possibly a fragment from a broken core.

Table 8: Maximum artefact size (mm) (Maximum size in any dimension)

Site & square	0-9.5	10-19.5	20-29.5	30-39.5	40-49.5	50-59.5	60-69.5	70-79.5	80-89.5	90-99.5	100 – 109.5	Total artefacts
Haynes 4A		17	16	4	4					1	1	43
Haynes 4B	4	45	30	14	3							96
Noala 1A		6	6	3		2	1		1			19
Total	4	68	52	21	7	2	1	0	1	1	1	158

Table 9: Maximum artefact size and general fauna (mm) Assemblages arranged from oldest on the bottom to youngest on the top.)

Site	Fauna	0-9.5	10-19.5	20-29.5	30-39.5	40-49.5	50-59.5	60-69.5	70-79.5	80-89.5	90-99.5	100 – 109.5	Total artefacts
Haynes	Mostly marine	3	30	24	14	5						1 core	77
Haynes	Marine & mammal	1	32	22	4	2					1 core		62
Noala	Marine & mammal		1		1			1		1 core			4
Noala	Mammal		5	6	2		1 & 1 core						15

The sizes of flakes, broken flakes and the Haynes Cave spit 3 core, associated with marine-dominated faunas, are plotted in Figure 6. It is possible that the smaller flakes were struck from the core or other similar cores, but some of the artefacts are larger than the core scars. The range of artefact sizes associated with marine and mammal fauna, from the lower deposit of Haynes Cave are like the range of scar sizes borne by the core from spit 3 in Noala Cave (Figure 7). Flake scars on the single-platform core in that assemblage tend to be larger. Few artefacts were recovered from the lower deposit in Noala Cave 1A, associated with mammalian fauna (Figure 8), but the large flake from spit 7 is notably larger than any other. It is closest in size to the largest scar on the single platform core from Haynes Cave.

The range of sizes of flakes and broken flakes did not always match the range of core scar sizes with which they were associated. It is possible that the cores which were discarded at these sites were not necessarily the same cores that were flaked in the caves or from which these flakes were produced, elsewhere off-site.

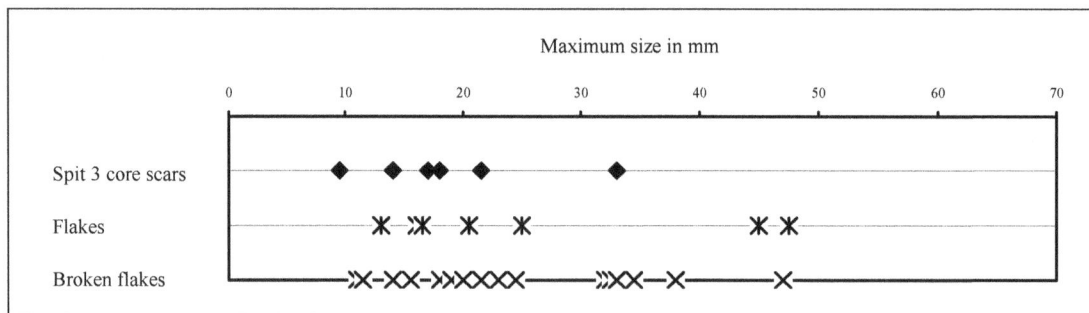

Figure 6: Size of core scars and debitage associated with marine-dominated fauna, Haynes Cave 4

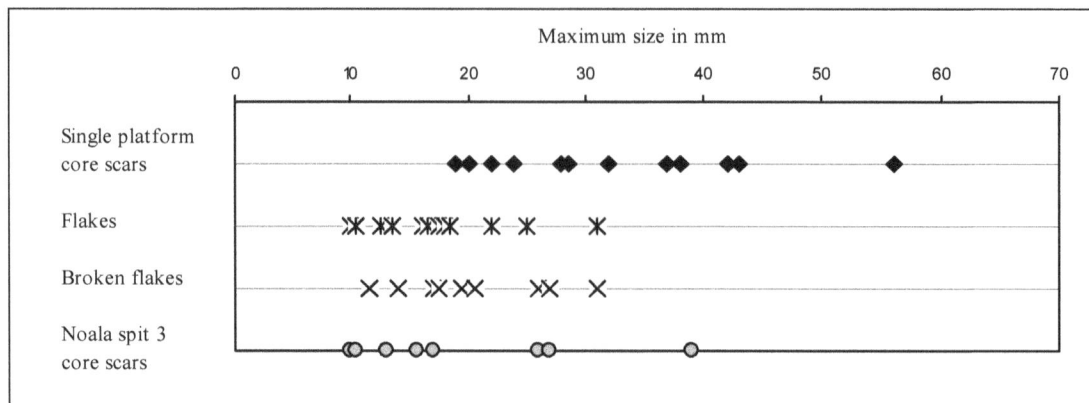

Figure 7: Size of core scars and debitage associated with marine and mammal fauna, Haynes Cave 4 and Noala Cave 1A

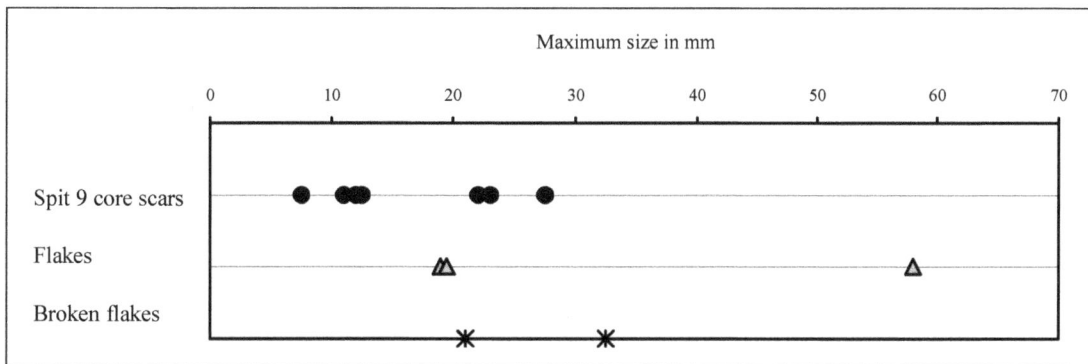

Figure 8: Size of core scars and debitage associated with mammal faunal remains, Noala Cave 1A

Table 10: Summary data on the size of core scars and debitage

Fauna	Site	Category	Number	Range of maximum size (mm)	Average size (mm)
Marine dominated	Haynes Cave 4	Spit 3 core scars	6	9.5 to 33	18.8
		Flakes	7	13 to 47.5	26.2
		Broken flakes	18	11 to 47	26.2
Marine & mammal	Haynes Cave 4	Spit 9 core scars	13	19 to 56	31.5
		Flakes	14	10 to 31	17.5
		Broken flakes	11	11.5 to 31	20.1
Marine & mammal	Noala Cave 1A	Spit 3 core scars	8	10 to 39	19.8
		Flakes	0		
		Broken flakes	0		
Mammal dominated	Noala Cave 1A	Spit 9 core scars	7	7.5 to 27.5	16.5
		Flakes	4	19 to 58	31.5
		Broken flakes	(2)	(21 to 32.5)	(26.8)

6.4 Flake shape

Core scars and flakes from the upper deposits of Haynes Cave 4, which are dominated by marine fauna, tend to be fairly squat, being as-wide-as-long or wider-than-long (Table 11, Figure 9). The range of elongation, and the average elongation for both data sets, is very similar. While a couple of flakes are larger than the core scars measured, they are essentially of the same range of shapes. This suggests that another larger but otherwise similar core may have been flaked at the site.

Core scars and flakes from the lower deposits of Haynes Cave 4, associated with both marine and mammal fauna, vary widely in shape (Table 11, Figure 10). Notable, however, is the presence of elongate flakes and core scars, which do not occur in the smaller sample from the upper assemblage associated with marine-dominated fauna. The presence of these slender artefacts and scars results in a higher elongation index than the later

assemblage. The larger, single-platform core shows a particularly wide range of flake scar shapes.

The apparent differences in flake elongation between the upper and lower deposits in Haynes Cave was assessed by assigning those with elongation indices of less than 0.9 and those with elongation indices of 0.9 or greater into two groups. The differing distributions are significant at the 0.05 level for a two-tailed test ($X^2 = 4.946$, $df = 1$).

The very small sample of flakes and core scars from the lower deposit at Noala Cave 1A, associated with mammal fauna, has a similar average elongation as the assemblage associated with marine and mammal fauna from the lower deposits of Haynes Caves, but notable in Figure 11 is the restricted range of flake and scar widths, giving the scatter plot a flatter appearance.

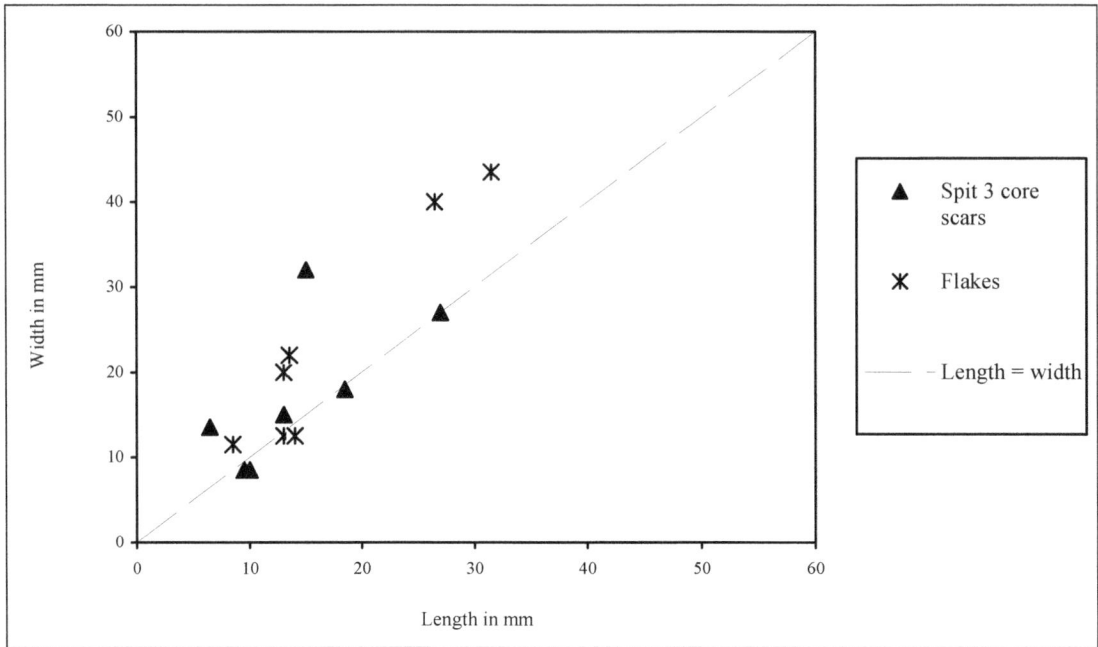

Figure 9: Shape of core scars and debitage associated with marine-dominated fauna, Haynes Cave 4

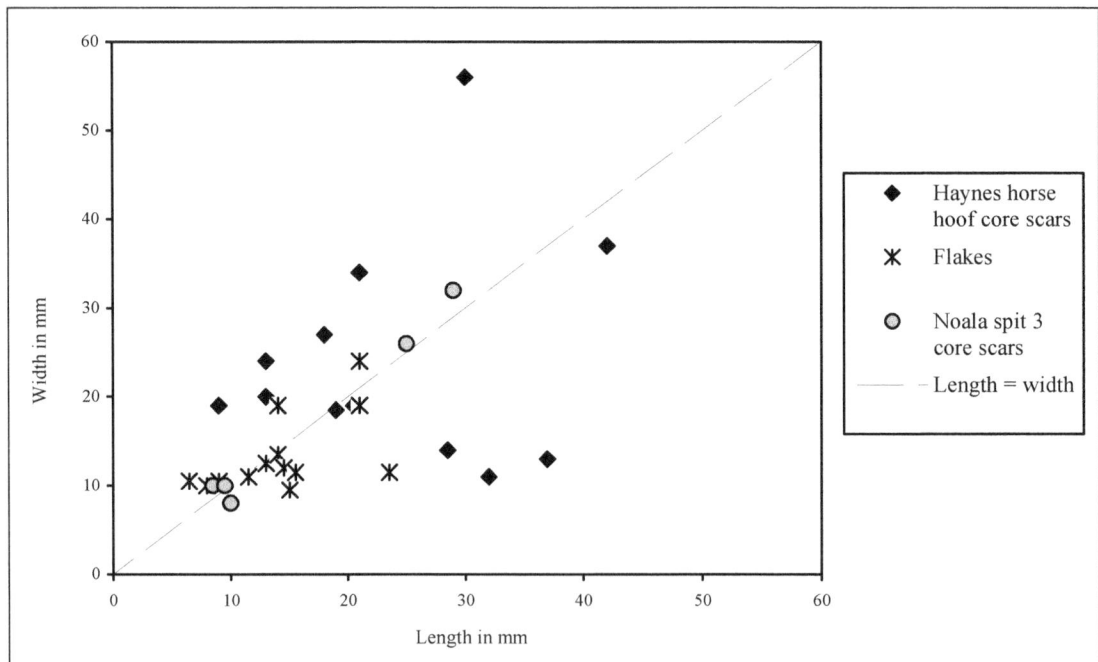

Figure 10: Shape of core scars and debitage associated with marine and mammalian fauna, Haynes Cave 4 and Noala Cave 1A

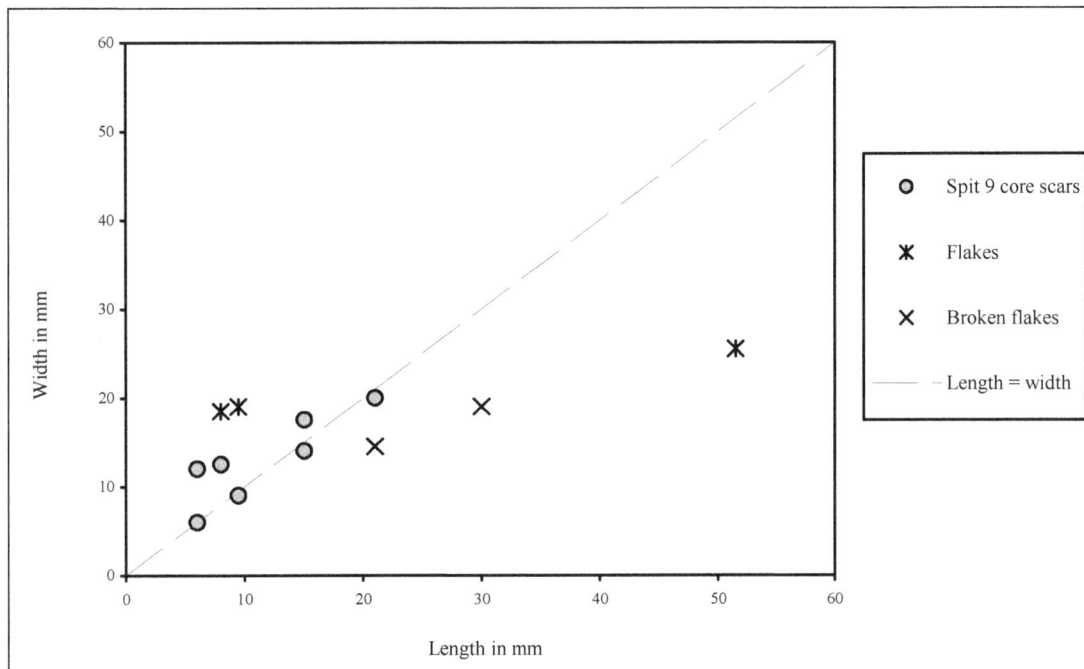

Figure 11: Shape of core scars and debitage associated with mammalian faunal remains, Noala Cave 1A

Table 11: Summary data on the shape of core scars and debitage (Elongation index is length divided by width; <1.0 flakes and scars are wider than long, >1.0 flakes and scars are longer than wide; ≥ 2.0 flakes and scars are elongate.)

Fauna	Site	Category	Number	Range of elongation	Average elongation
Marine dominated	Haynes Cave 4	Spit 3 core scars	7	0.5 to 1.2	0.8
		Flakes	7	0.6 to 1.1	0.7
Marine & mammal	Haynes Cave 4	Spit 9 core scars	12	0.4 to 2.9	1.0
		Flakes	14	0.6 to 2.0	1.1
Marine & mammal	Noala Cave 1A	Spit 3 core scars	5	0.9 to 1.3	1.0
		Flakes	0		
Mammal dominated	Noala Cave 1A	Spit 9 core scars	7	0.5 to 1.1	0.9
		Flakes	4	0.5 to 2.0	1.1
		Broken flakes	2	1.5 to 1.6	1.5

6.5 *Debitage* platform surfaces

Platforms (or part thereof) were present on 42 flakes and proximal broken flakes (Table 12). Most consist of smooth plain surfaces (67%). A few platforms have remnant flake margins on their surfaces, indicating that they were struck from rotated cores. A few others are focal (14%) (i.e. less than or equal to twice the area of the ringcrack), indicating overhang removal or precision in flaking (Hiscock 1986). Only two platforms have the initiation points of flake scars on their surfaces. Some other flakes and flake fragments have evidence for crushed or partly-crushed platforms.

The flake of unidentified stone from Noala Cave 1A spit 7 has a ridged platform (i.e. a remnant margin of flake on the platform) indicating that the flake was struck from a rotated core.

Platform surfaces show no substantial correlation with the different dietary phases through time.

Table 12: Platform surfaces

Fauna	Site & square	Plain	Ridged	Scar	Scar & ridge	Focal	Total identified platforms	Crushed	Indeter-minate
Marine dominated	Haynes Cave 4	13	2	1		2	18	3	7
Marine & mammal	Haynes Cave 4	11	3		1	4	19	6	4
Mammal dominated	Noala Cave 1A	4	1				5	4	
Total		28	6	1	1	6	42	13	11

6.6 Platform angles

The platform angles on the cores from Noala and Haynes Caves vary (Table 13). The lowest angle is that of the bifacial platform on the core from Haynes Cave 4A spit 3. The highest angles are on the single platform ('horse hoof') core from spit 9. Modal angle on debitage artefacts is $60° - 69°$, followed by $70° - 99°$ (Table 14).

Most of the *debitage* appear to have been struck from cores with platform angles a little lower than that of the single platform core from spit 9; and high platform angles on this core may have been a contributing factor to its discard. Some *debitage* were struck from cores with low platform angles, of less than $50°$, as shown in the instance of the bifacial core from Haynes Cave 4A spit 3.

There was little to distinguish between the assemblages associated with different faunal suites through time (Table 14).

The only published Australian analysis of platform angles of which I am aware is that relating to assemblages from the Hunter Valley of NSW (Hiscock 1986). In that study more than 20% of a Pre-Bondaian[2] flake assemblage had platform angles of less than $60°$ while fewer than 16% of flakes in Bondaian assemblages from the late Holocene had angles of less than $60°$ (Hiscock 1986:45). The Montebello assemblages have a comparable frequency of 22% of artefacts with angles less than $60°$.

6.7 Terminations

Scars on the cores are dominated by step and hinge terminations, with several scars on the core from Noala Cave spit 9 having plunging terminations (Table 15). Overall, feather terminations are most frequent on *debitage*, with half (50%) of artefacts displaying this type of termination (Table 16). Feather terminations appear a little more frequently on artefacts associated with marine fauna (60% of terminations) than in other assemblages (44% and 40%) but this could have resulted from random factors due to the small total numbers of artefacts. A chi-square analysis comparing feather terminations alone against all other terminations did not indicate a significant difference ($X^2 = 2.044$, $df = 3$).

Hinge and step terminations, alone or in combination, are also quite frequent (46%), potentially making the detachment of further flakes difficult. However, very few flakes have plunging terminations, indicating that few cores were reset within the caves by the complete removal of previous flake scars.

Within the samples Haynes Cave 4B has a higher frequency (57%) of feather terminations than does the assemblages from Noala Cave and Haynes Cave 4A where some 40% of artefacts have feather terminations. This is further evidence for intra-site variation within Haynes Cave.

6.8 Discussion: technical aspects of flaking

It was noted above (section 6.1) that the four cores recovered from Noala and Haynes Caves illustrate similar approaches to the procurement and reduction of calcarenite. Essentially, it appears that natural pieces of tabular stone were selected. These pieces may have originally been more than 10cm in size, with weights varying up to several hundred grams. The pieces selected appear not to have been particularly thick, with three of the four cores measuring 3cm or less in their third dimension, while the single-platform core is thicker, at nearly 5cm. These stone pieces were mainly flaked by unifacial (unidirectional) flaking, removing flakes from the shortest dimensions of the cores. Flaking was predominantly across the bedding plane of the stone, with core scars and most flaked debitage revealing the horizontal bands of the stone. Apart from the core from Noala Cave 1A spit 9 there was little attempt to remove longer flakes by flaking down the longer dimensions of the cores. Only a few pieces of *debitage* show the natural stone bands running parallel to flake length; an unusually large flake from Noala Cave 1A being one such exception.

In other regions of Australia where stone with strong horizontal bedding planes or structures has been flaked (e.g. fossil siliceous wood) narrow artefacts are sometimes flaked parallel to the inherent structure to produce blades (e.g. on the Cumberland Plain in Western Sydney). If the intention was not to produce narrow flakes or blades, flaking down the short axis of tabular banded stone allows for the production of numerous small flakes. However, such flaking can result in step and hinge terminations and higher platform angles. On a thinner core, platform angles and remnant step and hinge terminations can be removed by striking a plunging flake – that is, by striking a flake which removes part of the face and bottom of the core. This approach is seen on the core from Noala Cave 1A spit 9. Platform angles on the single-platform core were higher

---

[2] In that region the assemblage pre-dating the phase equivalent to the "Small Tool Tradition" which includes backed artefacts and faceted platforms.

than those shown by most of the *debitage*. On such a thick core it may have been more difficult to remove plunging terminated flakes.

Analysis of the maximum size and shapes of core scars indicates that most of the flakes which were struck from the three thinner cores tended to be wider than long in shape with elongation indices ranging between 0.5 and 1.3. The single-platform core included larger flake scars and scars which varied greatly in shape, with elongation indices ranging between 0.5 to 2.9.

It is possible that the thinner cores may have enabled production of smaller but more standardized flakes, while the single-platform core may have allowed production of a wider range of shapes and sizes, potentially useful if the nature of tool tasks varied, requiring different shaped edges or tool mass.

The core from Haynes Cave 4A spit 3 is unusual in the assemblages as it has a low angle bifacially flaked platform. This bifacial platform appears worn – it may have been used as a tool and/or the artefact may have been curated for some time. The artefact appears to have then broken, bisecting the bifacial platform. The core was then flaked unifacially, with flakes removed from the shorter axis of the core. The (remnant) low angle bifacial platform on this core demonstrates a different approach to stone reduction from that shown in the range of other artefacts recovered from Noala and Haynes Caves. Few pieces of *debitage* have such low platform angles, and few have scarred platforms.

A large flake, of a size unusual for the assemblages analysed here, was recovered from Noala Cave 1A spit 7. This flake is 58mm in size, elongate in shape and has a plunging termination. The banding pattern in the stone indicates that it was struck parallel with bedding planes, and possibly from the long axis of a core. The oriented length of the flake is 52mm, indicating that the core was this long. The size and shape of this flake falls outside the ranges shown by other artefacts and core scars. It is likely to have been struck from a type or size of core not recovered during the present investigations.

Table 13: Core platform angles

Fauna	Site	Spit	Angles	Comments
Marine dominated	Haynes 4A	3	45°	Bifacial platform
			68°	Unifacial platform
Marine & mammal	Haynes 4A	9	80° – 90°	Horse hoof core
Marine & mammal	Noala 1A	3	58° – 84°	
Mammal dominated	Noala 1A	9	60° – 87°	

Table 14: Debitage platform angles

Fauna	Site & square	Range	Mean	30° – 39°	40° – 49°	50° – 59°	60° – 69°	70° – 79°	80° – 89°	90° – 99°	Total
Marine dominated	Haynes Cave 4	37° – 87°	69°	1		5	9	4	6	2	27
Marine & mammal	Haynes Cave 4	34° – 95°	66°	1	2	4	8	8	3		26
Mammal dominated	Noala Cave 1A	66° – 73°	69°				3	2			5
Total				2	2	9	20	14	9	2	58

Table 15: Core scar terminations

Site	Spit	Feather	Feather & hinge	Hinge	Hinge & step	Step	Feather & step	Plunging	Total
Haynes 4A	3	3	1	2		1	1		8
Haynes 4A	9	4	3	3	1	2			13
Noala 1A	3	1	1	2		2		2	8
Noala 1A	9			2		4	1	6	13
Total		8	5	9	1	9	2	8	42

Table 16: Debitage terminations

Fauna	Site	Feather	Feather & hinge	Hinge	Hinge & step	Step	Feather & step	Plunging	Total
Marine dominated	Haynes 4	18	4	5	1	1		1	30
Marine & mammal	Haynes 4	16	6	6	4	2	1	1	36
Mammal dominated	Noala 1A	4	2	2	1			1	10
Total		38	12	13	6	3	1	3	76

Some of the broken flakes from Haynes Cave show that some artefacts split along the bedding planes, breaking flakes transversely. Experimental knapping may shed light on whether this kind of breakage resulted from inherent weaknesses in the stone, or whether some of this breakage occurred after discard, during the intervening millennia.

6.9 Discussion: technical aspects of flaking and changing economic base

The Montebello Islands provide an uncommon opportunity to investigate technical aspects of flaking in relation to demonstrable changes in the broad resource base over time. The samples of lithics are small, and larger numbers would allow more confidence in the analysis; but there are some interesting correlations, summarised below, between technical aspects of the assemblage (core size, artefact size and flake shape) and the differing suites of faunal remains. The variables of type of flake platform, platform angles and type of termination show little variation.

Assemblage associated with mammal faunal remains, Noala Cave 1A. Artefact discard rates were relatively low in the lower deposit of Noala Cave 1A when the lithics were associated with fauna dominated by mammalian remains. The few artefacts recovered (n=15) vary in size, and include the largest flake analysed for this report. There is some suggestion that flakes tend to be less wide than in later assemblages but a larger sample size would be required to allow some confidence in this observation.

Assemblages associated with marine and mammal faunal remains. Few artefacts were recovered from the upper deposit at Noala Cave 1A, when marine fauna appears in the dietary remains. However, larger numbers of artefacts were recovered from Haynes Cave when both mammalian and marine faunal remains were common. The presence of more numerous flakes in the 1cm to 3cm size ranges may indicate an increase in on-site flaking. Flakes and scars on the single-platform core from Haynes Cave vary in shape, from twice-as-wide-as-long to almost three-times-as-long-as-wide. In a mixed economy, variation in flake shape may have been advantageous, given a potentially wider range of tool task requirements. It is possible that there was lower residential mobility if proximal resources were more plentiful and localised; larger core size and wider variation in flake shape could have been related to lower mobility and fewer constraints posed by portability (Mulvaney 1975:73).

Assemblage associated with marine faunal remains. The lithic assemblage from the upper unit of Haynes Cave, when marine faunal remains were dominant also includes artefacts less than 3cm in size, as well as a higher frequency of artefacts in the 3cm to 5cm size range. The scars on the core associated with this assemblage did not match the range of artefact sizes, indicating that at least some of the artefacts were struck from a larger core. The recovered flakes and the core scars tend to be wider-than-long in shape, and the larger flakes are of similar shapes to the smaller flakes and scars. It is possible that relatively long narrow flakes were not necessary for tasks related to marine resource exploitation.

7. COMPARISONS WITH CAPE RANGE LITHICS

Other lithic assemblages from several sites on the Cape Range Peninsula, about 250km south-south-west of Montebello Islands have been analysed (Morse 1993, Przywolnik 2002). Cape Range is the central limestone spine of the peninsula (Przywolnik 2002:33), not entirely unlike the geology of Barrow and Montebello Islands.

Lithic raw materials occur naturally in the Cape Range region, with silcrete, sandstone, siltstone and "partially silicified limestone" occurring as pebbles (Morse 1993:64) and sometimes quarried from outcrops (Przywolnik 2002:66). Chalcedony is exotic (Morse 1993:64). The lithologies utilised for stone artefacts vary but silcrete and "partially silicified limestone" tend to be dominant (Morse 1993, Przywolnik 2002) – that is, locally available lithologies were preferred. Both Morse and Przywolnik noted a tendency towards increased use of fine-grained lithologies for mid-to-late Holocene assemblages compared to earlier assemblages.

Data on modified artefact types are reported on (Table 17). Generally, retouched and formal tools are more frequent amongst recent assemblages than those of the Early Holocene or Pleistocene (Morse 1993, Przywolnik 2002), although the frequencies vary between sites. Small numbers of adzes, adze slugs and backed artefacts occur in the assemblages from the mid- to late Holocene. The absence of these types from the Noala Cave and Haynes Cave sites is consistent with their earlier dating.

A few fragments of grindstones and/or mullers (topstones) occur in shelters C99 Layers 1 to 3 and Jansz Layer 3. Layer 3 in these sites is dated to 7,700 BP and 12,400 cal BP, respectively (Przywolnik 2002:164, 217).

The frequencies of cores and core fragments vary between the Cape Range sites over time. In the Early Holocene and Pleistocene assemblages cores comprise 15% of the aassemblage at Mandu Mandu and only 1.4% at C99 (Morse 1993, Przywolnik 2002). At least two horse hoof cores were reported from Mandu Mandu Creek Unit 2 (Morse 1993:170). Core discard rates appear to have been highly variable.

At Mandu Mandu shelter core sizes vary over time, with the older Unit 2 cores being notably larger than Unit 1 cores (132g/core in Unit 2 compared to 23g/core in Unit 1) (Morse 1993:169). Drawings of the Mandu Mandu horse hoof cores indicate that they are about 7cm to 10cm across (Morse 1993:170) – sizes not unlike that from Haynes Cave. Indeed, three of the four cores from Noala and Haynes Caves are within the size range of the Mandu Mandu horse hoof cores, and the four cores have a relatively high average weight of 218g/core.

Summary data on the sizes and weights of flakes are presented by Morse (1993). The average length and average weight of flakes from Mandu Mandu show change over time with the more recent Unit 1 flakes significantly smaller and lighter than the earlier Unit 2 flakes (Morse 1993:167). Yardie Well appears to show a similar trend (Table 18).

Only three complete flakes were recovered from Noala Cave. A larger sample of 21 complete flakes was recovered from Haynes Cave and have a smaller average length and a notably less heavy than the Cape Range assemblages. Smaller flake size could be related to the transport of stone from Barrow Island – the nearest known source of calcarenite to the Montebello Islands.

8. SUMMARY AND CONCLUSIONS

Artefact numbers. There is considerable variation in the numbers of artefacts recovered, between and within the two sites. At Noala Cave 1A only a small number of artefacts were recovered with more (n = 15) occurring in the lower unit than in the upper one (n = 4). Most of the Noala Cave assemblage appears to be associated with occupations marked by exploitation of land mammals.

In Haynes Cave 4A small numbers of artefacts were found in all spits through the deposit. In 4B however, twice as many artefacts were recovered, and these tended to occur in "bands" with higher concentrations in spits 1, 3 and 4, and 6 to 8 and only a few in spits 2 and 5.

Table 17: Cape Range sites – summary distribution of artefact types by % frequency

Site	Cores & fragments	Retouch/ formal	Notable types	Aprox. age estimate BP	Total (n)	Reference
Shelters: Mandu Mandu Unit 1	13.7	6.2	Adzes, ground	300 – 5,600	916	Morse 1993:166
Mandu Mandu Unit 2	14.7	2.7	Notched tools, horse hoof core	22,000 – 35,000	691	
Pilgonaman Unit 1	9.9	3.2	Adzes & slugs	500	471	Morse 1993:223
Pilgonaman Unit 2	10.9	2.4		5,000 to 32,000	382	
Yardie Well Unit 1	8.9	5.1	Backed & adzes	< 5,700 ?	78	Morse 1993:254
Yardie Well Unit 2	5.5	0		7,000 – 10,5000	72	
C99 Layer 1	2.2	0.8	Burren adze, ground	800 - 2,000	993	Przywolnik 2002:199, 200
C99 Layer 2	1.7	0.6	Ground pieces	50 – 2,400	648	
C99 Layer 3	1.4	0.6	Burren adze, ground	7,700 – 9,000	855	
C99 Layer 4	2.4	0		21,000 – 34,200	42	
Jansz Layer 1	3.6	2.4	Backed	0 – 500	83	Przywolnik 2002:241, 243
Jansz Layer 2	1.2	1.7	Tula	300 – 600	353	
Jansz Layer 3	4.1	0	Ground	9,700 – 12,400	469	
Jansz Layer 4	4.0	0		30,400 – 36,300	299	
Noala Cave	10.5	0			19	
Haynes Cave	1.4	0.7	Horse hoof core		139	

Table 18: Average size of flakes from Montebello and Cape Range (Length measured perpendicular through the PFA)

Assemblage	Mean length (mm)	Mean weight (g)	Total (n)	Reference
Mandu Mandu Unit 1	19.7	2.39	259	Morse (1993:168)
Mandu Mandu Unit 2	22.3	3.96	176	Morse (1993:168)
Yardie Well Unit 1	16.7	3.42	17	Morse (1993:256)
Yardie Well Unit 2	25.1	8.57	30	Morse (1993:256)
Noala Cave	(23.0)	(10.3)	(3)	
Haynes Cave	15.0	1.6	21	

Lithologies. The lithics are predominantly of calcarenite; those from Noala Cave are predominantly of a fine grained stone, while those from Haynes Cave are a little coarser with bedding planes and horizontal banding evident in the stone. Three small artefacts of high quality quartz were also recovered; two from Noala Cave and one from Haynes Cave – the latter being a fragment of a retouched tool. Research carried out by MARP indicates that quartz does not outcrop in the Montebello or Barrow Islands and must have originated from a mainland supply zone. Another artefact from Noala Cave comprises a grey stone with cortex on its dorsal surface and it too may have originated from mainland sources. These four artefacts of exotic stone came from the lower spits in their squares (Noala Cave spits 7 and 10, Haynes Cave 4B spit 8) – levels which date to the time when the Montebello Islands were connected to the mainland.

Dating and analysis of faunal remains indicate that lithics from the lower deposit at Noala Cave relate to a time of climatic amelioration (12,400 to 10,000 BP) when the coast was still relatively distant. At this time the visitors to Noala Cave may have had access to a fine-grained calcarenite source, together with high quality quartz and other lithic materials procured or exchanged from (now) mainland sources. Artefacts of these exotic lithologies make up 20% of the assemblage from the lower deposits of Noala Cave. The quartz retouched fragment from the deepest spit in Haynes Cave 4B makes up less than 2% of its associated assemblage, indicating very limited contact (if any) with the mainland at this time (it may have been curated/re-used from an earlier discard).

Range of variation. Within Haynes Cave some variation is evident in the lithics between 4A and 4B with 4B having larger numbers of artefacts, higher rates of artefact breakage, no cores (while 4A had fewer artefacts but two cores), higher frequency of artefacts in the 1-2cm size range, and a higher frequency of artefacts with feather terminations. Given the presence of intra-site variation it is possible that other locations within the caves could provide assemblages with different technical characteristics.

Predominant knapping strategy. Technical analysis of the cores and *debitage* indicates that the main knapping strategy involved flaking tabular pieces of banded calcarenite, generally less than 3cm thick. Flaking the short dimension of these cores may have enabled production of numerous, fairly small and squat flakes, with platform angles, and remnant step and hinge terminations, correctable by the removal of plunging flakes. Flaking could be carried out

simply and easily to meet task requirements as they arose. The thicker single-platform ('horse hoof') core allowed for the production of larger flakes and flakes varying in shape, potentially able to meet a more varied range of tasks. However, the advantage of the single-platform core may have been off-set by the increased difficulty of resetting the platform and removing unwanted step and hinge terminations by striking plunging flakes. Other reduction strategies are indicated by the bifacial core, with sufficient mass and low platform angle potentially suited to use as a heavy duty tool. The relatively large flake in Noala Cave 1A spit 7 also appears to have derived from a different reduction strategy.

Technical attributes and use of economic resources. There are some tantalizing correlations between changes in faunal remains and some technical aspects of the lithic assemblages through time.

The few artefacts (n=15) associated with predominantly mammalian faunal remains at Noala Cave suggest that artefact discard during this phase may have been limited, perhaps to conserve stone. The few flakes and core scars are of limited width. This assemblage also included the smallest core recovered. It is possible that residential mobility may have been higher at this time than later.

Higher numbers of artefacts were recovered from Haynes Cave when both mammalian and marine faunal remains are present, with considerably more small flakes indicating an increase in on-site flaking. Two cores – one from the upper deposit of Noala Cave 1A and a horse hoof core from the deeper deposit of Haynes Cave - are the heaviest cores recovered. Flakes, and scars on the horse hoof core, vary greatly in shape. In a mixed economy, variation in flake shape may have been advantageous, given a potentially wide range of required tasks. Larger core size and wider variation in flake shapes could have been related to lower mobility and fewer constraints posed by portability (*sensu* Mulvaney 1975:73).

The lithic assemblage from the upper portion of Haynes Cave, when marine faunal remains were dominant also includes artefacts less than 3cm in size, as well as a higher frequency of artefacts in the 3cm to 5cm size range. The recovered flakes and the core scars tend to be wider-than-long in shape. It is possible that relatively long narrow flakes were not necessary for tasks related to the marine-based economy. The core from spit 3 at Haynes Cave is the largest core recovered (as measured in mm) but notably lighter than the two from the earlier phase.

Comparisons with Cape Range artefact assemblages. Detailed comparisons with assemblages from Cape Range sites are restricted by a lack of technical data. Formal artefact types such as tulas and backed artefacts were sometimes found within Late Holocene assemblages from Cape Range. None were recovered from the small number of artefacts recovered from the Montebello Islands – and nor would they be expected given the earlier dates for the Montebello sites. At least two large single-platform (horse hoof cores) were reported for Mandu Mandu Unit 2, and from the drawings (Morse 1993:170) appear to be of dimensions similar to that from Haynes Cave. Data on the average lengths and weights of flakes suggests that the sample from Haynes Cave (although consisting of only 21 flakes) may have been smaller than those from the Mandu Mandu and Yardie Well sites.

Small sample size. The analyses and interpretations made in this report must be regarded as provisional, given the small numbers of artefacts in the assemblages. While interpretations of change in artefact size and flake shape appear to be based on statistically significant data, it is possible that a larger numbers of artefacts would illustrate greater variation in technical attributes (after Leonard and Jones 1989).

REFERENCES

Baker, N. B. 1992 Evidence from the analysis of cores. (In) Narama salvage project, Lower Bayswater Creek, Hunter Valley, NSW, pp. 10-42. Vol. 4: Technological studies. Unpublished consulting report prepared by Brayshaw McDonald Pty Ltd for Envirosciences and Narama Joint Venture.

Byrne, D. 1980 Dynamics of dispersion: the place of silcrete in archaeological assemblages from the lower Murchison, Western Australia *Archaeology and Physical Anthropology in Oceania* 15:110-119.

Clegg, F. 1990 *Simple statistics: a course book for the social sciences* Cambridge University Press, Cambridge.

Cotterell, B. and Kamminga, J. 1987 The formation of flakes *American Antiquity* 52(4): 675-708.

Hiscock, P. 1986 Technological change in the Hunter River valley and its implications for the interpretation of late Holocene change in Australia *Archaeology in Oceania* 21(1): 40-50.

Hiscock, P. 1988 Prehistoric settlement patterns and artefact manufacture at Lawn Hill, North-west Queensland. Unpublished Ph.D. thesis, Department of Anthropology and Sociology, University of Queensland.

Hiscock, P. and Attenbrow, V. 2002 Morphological and reduction continuums in Eastern Austalia: measurement and implications at Capertee 3. *Tempus* 7: 167-174.

Holdaway, S. 1995 Stone artefacts and the transition. (In) J. Allen and J. F. O'Connell (eds) *Transitions: Pleistocene to Holocene in Australia and Papua New Guinea*, pp.784-797. *Antiquity* 69 (special number 265).

Holdaway, S. and Stern, N. 2004 *A record in stone: the study of Australia's flaked stone artefacts* Museum Victoria and Aboriginal Studies Press.

Jones, R. (ed) 1985 *Archaeological research in Kakadu National Park*. Special publication 13, Australian National Parks and Wildlife Service and Prehistory Pacific Studies, Australian National University, Canberra.

Kuhn, S. L. 1994 A formal approach to the design and assembly of mobile toolkits. *American Antiquity* 59(3): 426-442.

Leonard, R.D. and Jones, G.T. 1989 *Quantifying diversity in archaeology* Cambridge University Press, Cambridge.

Lourandos, H. 1997 *Continent of hunter-gatherers: new perspectives in Australian prehistory* Cambridge University Press, Cambridge.

McNiven, I. J. 1994 Technological organization and settlement in southwest Tasmania after the glacial maximum. *Antiquity* 68:75-82

Morse, K. 1993 West side story: towards a prehistory of the Cape Range Peninsula, Western Australia. Unpublished Ph.D. thesis, Centre for Archaeology, University of Western Australia

Mulvaney, D. J. 1975 *The Prehistory of Australia* Penguin Books, Middlesex.

Mulvaney, J. and Kamminga, J. 1999 *Prehistory of Australia*. Allen & Unwin, Sydney.

Przywolnik, K. 2002 Patterns of occupation in Cape Range Peninsula (WA) over the last 36,000 years. Unpublished Ph.D. thesis, Centre for Archaeology, University of Western Australia.

Shott, M. 1986 Technological organization and settlement mobility: an ethnographic example. *Journal of Anthropological Research* 42(1): 15-51.

Speth, P. 1972 Mechanical basis of percussion flaking. *American Antiquity* 37(1): 34-60.

Witter, D. C. 1992 Regions and Resources. Unpublished Ph.D. thesis, Research School of Pacific and Asian Studies, Department of Prehistory, Australian National University, Canberra.

Wright, R. V. S. 1983 Stone implements. (In) G. Connah (ed) *Australian field Archaeology: a guide to techniques*, pp.118-125. Australian Institute of Aboriginal Studies, Canberra.

www.ingramcontent.com/pod-product-compliance
Lightning Source LLC
Chambersburg PA
CBHW061302270326
41932CB00029B/3442